INSTITUTION OF CIVIL ENGINEERS

Wetland management

Proceedings of the international conference organized by the Institution of Civil Engineers and held in London on 2–3 June 1994

Edited by R. A. Falconer and P. Goodwin

Organizing Committees

UK: Professor R. A. Falconer, University of Bradford, (Chairman); EurIng C. J. A. Binnie, W S Atkins and Partners; P. S. Lee, Mott MacDonald International; Professor J. S. Pethick, Institute of Estuarine and Coastal Studies; Ir H. Smit, Rijkswaterstaat; and Dr C. Swinnerton, National Rivers Authority

Overseas: Dr P. Goodwin, Philip Williams & Associates, USA (Chairman); Professor K. Nakatsuji, Osaka University, Japan; Dr Ong Jin Eong, Universiti Sains Malaysia; and Dr S. B. Reed, Hong Kong Government Environmental Protection Department

Published on behalf of the organizers by Thomas Telford Services Ltd, Thomas Telford House, 1 Heron Quay, London E14 4JD

First published 1994

Distributors for Thomas Telford are
USA: American Society of Civil Engineers, Publications Sales Department, 345 East 47th Street, New York, NY 10017-2398
Japan: Maruzen Co. Ltd, Book Department, 3–10 Nihonbashi 2-chome, Chuo-ku, Tokyo 103
Australia: DA Books & Journals, 648 Whitehorse Road, Mitcham 3132, Victoria

A CIP catalogue record for this publication is available from the British Library.

Classification
Availability: Unrestricted
Content: Collected papers
Status: Authors' opinion
User: Engineers, environmental planners, biologists, ecologists, hydrologists and geomorphologists

ISBN 0 7277 1994 7

© The Authors and the Institution of Civil Engineers, 1994, unless otherwise stated.

All rights, including translation, reserved. Except for fair copying, no part of this publication may be reproduced, stored in a retrieval system or transmitted in any form or by any means electronic, mechanical, photocopying, recording or otherwise, without the prior written permission of the Publications Manager, Publications Division, Thomas Telford Services Ltd, 1 Heron Quay, London E14 4JD.

Papers or other contributions and the statements made or opinions expressed therein are published on the understanding that the author of the contribution is solely responsible for the opinons expressed in it and that its publication does not necessarily imply that such statements and or opinions are or reflect the views or opinions of the organizers or publishers.

Printed and bound in Great Britain by Cromwell Press, Melksham, Wilts.

Contents

From reclamation to restoration – changing perspectives in wetland management. P. B. WILLIAMS	1
Lessons learned in wetlands restoration and enhancement. P. GRENELL	7
Wetland archaeology and wetland management. B. J. COLES	20
Human pressures on natural wetlands: sustainable use or sustained abuse? R. S. K. BUISSON and P. BRADLEY	35
Pressures on wetlands. J. GARDINER	47
Estuaries and wetlands: function and form. J. PETHICK	75
Somerset levels and moors water level management and nature conservation strategy. K. W. TATEM and I. D. STURDY	88
The Rhine-Meusa Delta: ecological impacts of enclosure and prospects for estuary restoration. H. SMIT, R. SMITS and H. COOPS	106
Numerical modelling of hydrodynamic and water quality processes in an enclosed tidal wetland. R. A. FALCONER and S. Q. LIU	119
Physical processes in tidal wetland restoration. P. GOODWIN	130
An integrated eco and hydrodynamic model for prediction of wetland regime in the Danubian Lowland. J. C. REFSGAARD, K. HAVNO and J. K. JENSEN	143
The role of computational models in wetland management – a case study from Bangladesh. C. E. REEVE and N. WALMSLEY	156
Estuarine management: the advantages of an integrated policy analysis approach. B. KARSSEN, M. W. M. KUYPER, M. MARCHAND, A. ROELFZEMA and M. VIS	167
Halting and reversing wetland loss and degradation: a geographical perspective on hydrology and land use. G. E. HOLLIS	181
The value of regional perspective to coastal wetland restoration design. A. M. BARNETT, R. S. GROVE and S. L. BACZKOWSKI	197
Water treatment systems – the need for flexibility. R. I. COLLINSON, A. G. HOOPER and M. HANNAM	213

The wildlife value and potential of wetlands on industrial land. A. MERRITT	226
Management of the Essex saltmarshes for flood defence. D. J. LEGGETT and M. DIXON	232
Management initiatives for coastal wetlands. A. REYNOLDS and J. BROOKE	246
Seasonal changes of groundwater chemistry in Miyatoko mire. T. HIRATA, S. NOHARA, T. IWAKUMA, C. TANG and K. NAKATSUJI	260
The management of lagoons to conserve their natural heritage. J. G. MUNFORD and D. LAFFOLEY	270
An integrated nature conservation and development strategy for the Deep Bay wetlands, Yuen Long, Hong Kong. G. GRANT	283

From reclamation to restoration – changing perspectives in wetland management

EurIng P. B. WILLIAMS, BEng, PhD, CEng, MICE, PE, President, Philip Williams & Associates Ltd, San Francisco, and Honorary President, International Rivers Network

SYNOPSIS. The historical evolution of the way society has viewed wetlands is described. Human adaption to wetlands has been superseded by exploitation in the form of reclamation. Now the true costs of wetland destruction are being appreciated and the dominant paradigm of reclamation has been giving way to a new paradigm of conservation. However, the dramatic loss in wetland area that has already occurred and its impact at the local, regional and global scale has led to a new initiative for wetland restoration. Based on U.S. experience in the last 20 years, the key components of successful wetland restoration programmes are summarized.

ADAPTION TO WETLANDS

1. Human society has developed on the waters' edge. Our cities have grown up on river banks or shorelines. Our richest farmland has been taken from the valley bottoms and floodplains; our most productive fishing has always been in estuaries and coastal waters, fed by tidal marsh and mudflat. This intimacy between human history and wetlands is reflected in the history of the changing way society has perceived and managed its wetlands.

2. For most of human history we have made use of wetlands by *adaption*. Since time immemorial, local communities harvested and used the tremendous productivity of the resources afforded by wetlands in sustainable ways. European history provides many examples. In the U.K., water meadows had a special place in medieval agriculture and the wilderness of the fens created and supported its own unique and independent way of life. Similar adaptions have occurred all around the world from the seasonal floodplain agriculture of the Niger Valley, to the way of life of the "marsh" Arabs of Iraq, to the sophisticated farming methods on the floodplain lands in Bangladesh. In most of the world, until perhaps a generation ago, adaption was still the way most people lived with their wetlands.

RECLAMATION OF WETLANDS

3. Now this idea of adaption has crumbled before the onslaught of a newer idea derived from the industrial revolution's utilitarian views of nature that demanded resource exploitation for economic development. When applied to wetlands this ideas was implemented as *reclamation*, the transformation of wetland to dry land.

4. Now, as the whole world embraces the industrial revolution's definition of economic development, it also seeks to emulate the industrialized world's success in obliterating wetlands through reclamation. As reclamation is now the dominant paradigm around the world we should examine its rationale more closely. Conversion of wetland to dry land indubitably achieves a higher economic value for very specific purposes, such as farming or urban development, for particular interest groups in society such as farmers or land developers. These benefits are tangible, visible and usually quickly realized. However, as we have now learned, there are often huge costs involved, but these costs are far less tangible because they are environmental and cumulative and can take a long time to be translated into economic costs. These costs are usually not borne by the beneficiaries of reclamation, but by the taxpayer or society at large and, of course, the global ecosystem. It is perhaps in pursuit of an overriding rationalization that the advocates of resource exploitation succeeded in popularizing their activities as "reclaiming" rather than "claiming", implying that all wetlands were formerly dry lands and that mankind was merely correcting nature's mistake.

5. The transition from adaption to reclamation occurred when one particular interest group became politically powerful enough to expropriate a common resource. Many of us know that the history of England itself has been shaped by popular opposition to the large scale reclamation schemes promoted by Charles I and designed by expatriate technical experts like Cornelius Vermuyden. (English commoners who depended on wetlands for their livelihood took up arms against the king but were ultimately betrayed by their leader, Oliver Cromwell). Few know that there is similar opposition occurring right now to drainage and diversion schemes affecting wetlands around the world—whether the Danube floodplain forest of Slovakia or the Okavango Swamp of Botswana.

6. However it was not so much the plight of people affected directly by wetland reclamation that caused the next shift in our perception but a growing realization of the wholesale ecologic destruction caused by mankind's power to transform our wetland landscape. This destruction has been caused not only by draining, filling or embanking wetlands, but also by dams, diversions, increased sedimentation from deteriorating watersheds, pollution and introduced exotic species. The pace of destruction has accelerated in the last few decades with the advance of technology in improved pumps, earth moving equipment and the construction of ever larger dams.

7. Within the U.S. nearly 50% of all natural wetlands have been destroyed, but this understates the direct impact on some of the more important types of habitat. For example, the Mississippi floodplain forests have been reduced to only a few percent, California's wetlands have been reduced to 9%. The indirect impacts of these changes is apparent: reduction in waterfowl, reduction in coastal fisheries, continually increasing flood damages, and deteriorating water quality.

8. In the U.K. where 60% of wetlands have been destroyed and only 10 years ago wetlands were being reclaimed at a rate of 100,000 ha/yr, 6 wetland plant species have gone extinct since 1930, and calcareous fens and lowland peat bogs have been virtually eliminated.

9. Throughout Europe, the picture is the same: 40% of Brittany's coastal wetlands destroyed since 1960; 70% of Portugal's Algarve drained for agriculture; 18 out of 31 endangered bird species are dependant on shrinking areas of wetland habitat; and Europe's magnificent floodplain forests of the Rhine and Danube have been decimated.

10. In the rest of the world we belatedly have come to appreciate the importance of the 2% of the total earth's surface that comprise wetlands. For example, the critical role in protecting biodiversity by the floodplain forest of the Amazon, or the role of the Arctic tundra as a carbon sink that buffers the effects of global warming, or the importance of mangrove wetlands in supporting valuable coastal fisheries as well as protection against coastal flooding.

WETLAND CONSERVATION

11. Starting about 75 years ago in both the U.S. and U.K. a new movement developed to protect remaining areas of wetland habitat. This movement promoted the new of idea of wetland *conservation*, and was an expression of the beginnings of the environmental movement—then referred to as nature conservation.

12. In the U.S., wetlands had also found a new constituency amongst the politically powerful—the recreational duck hunters who were able to establish the first National Wildlife Refuge on the edge of the Florida Everglades in 1934. Similarly, in the U.K., the popular appreciation for the natural environment found political expression through County nature trusts like the Norfolk and Norwich Naturalists' Society, who were able to acquire wetlands for wild birds in the early 1920's.

13. In the U.S., the idea of wetland conservation gradually took root but was still a marginal activity until the growth of the environmental movement as a significant political force in the late 1960's. Aided by the popularization of Professor Eugene Odum's landmark research on ecology that used a Georgia salt marsh to demonstrate the interdependence of so many species, environmentalists were able to include statutory protection for some wetlands written into law as section 404 of the 1970 Clean Water Act. This protection is partial, and has taken a long time to interpret and implement. It has had to be buttressed with other legislation to protect endangered species—many of which are wetland dependant. But the end result is that in the U.S. there has been an important shift in public attitudes away from reclamation, and the dominant paradigm is now one of wetland conservation.

14. This change in attitude has also been occurring in Europe with attempts by the European Union to institute wetlands conservation policies. In the U.K., the battles over the drainage of the Halvergate marshes and the Somerset Levels in the early 1980's led to the loss of public support for agricultural subsidies for wetland reclamation. And now in Japan, opposition to the Chitose River diversion and the Nagara River Dam is causing the government to rethink its policies that encourage wetland reclamation and instead start to assist in conserving wetlands.

15. In the U.S., a further transformation in the perception of wetland management has now been taking place. As the idea of wetland conservation took hold, more research and public education has been focussed on wetland issues, leading to a deeper understanding of the importance of wetlands in the economy and the environment, and an appreciation for what has been lost worldwide, particularly in the last few decades.

WETLAND RESTORATION

16. This growing realization of the importance of the ecologic and economic damages caused by the destruction of wetlands led to a search for a way to heal some of this damage. This was the genesis of wetland *restoration*.

17. Starting about 20 years ago in the U.S., restoration was first seen primarily as compensation or mitigation for the destruction of wetlands elsewhere. The idea of mitigation explains the phrase "no net loss of wetlands," a policy frequently articulated by U.S. presidential candidates. However, by the early 1980's confusion over what "restoration" goals really were, and criticism of the lack of success of certain wetland restoration projects, had led many resource managers and environmentalists to question the value of restoration, if it opened the door to the destruction of pristine wetlands elsewhere. This skeptical view is clearly articulated on the global scene in WWF International's 1990 report *Wetland Conservation* that questions the value of wetland restoration and instead emphasizes the need for global wetland conservation.

18. In the U.S., within the last decade, new imperatives for wetland restoration have emerged dissociated with the idea of mitigation. The economic benefits of flood alleviation, water quality improvement, groundwater recharge, coastal defense, fishery production, tourism and property values associated with wetlands have become more and more apparent. We now understand that the long-term and cumulative ecologic damage caused by the destruction of wetlands due to draining, filling, embankments, reservoirs, diversions or groundwater pumping, has become yet another component of the global environmental crisis along with the destruction of forests or the ozone layer. We see this in the disappearance of amphibians, the decline of coastal fisheries or the imminent death of the last Everglades jaguar.

19. The rationale for wetland restoration as an active initiative rather than merely as a mitigation response is clearly articulated in the U.S. National Research Council report *The Restoration of Aquatic Ecosystems*, which calls for the restoration of 10 million acres in the U.S. in the next 20 years:

> "Without an active and ambitious restoration program in the United States, our swelling population and its increasing stresses on aquatic ecosystems will certainly reduce the quality of human life for present and future generations. By embarking now on a major national aquatic ecosystem restoration program, the United States can set an example of aquatic resource stewardship that ultimately will set an international example of environmental leadership.

20. Just in the last few years the impetus of wetland restoration in the U.S. has strengthened. For example in the aftermath of the 1993 Mississippi flood the U.S. Fish & Wildlife Service has developed a plan to restore 500 km of floodplain wetlands by removing levees along the Missouri River. In California 20% of the yield of the Bureau of Reclamation's Central Valley Project has been reallocated for restoring fish and wildlife. While the wholesale destruction of wetlands continues in most of the world, it is not just the U.S. that is changing direction. The Netherlands is experimenting with restoring polderland to wetlands on the Zuyder Zee and has announced its intent to ultimately restore approximately 200,000 ha of reclaimed wetland. In Germany, popular support is growing to restore the floodplain forests of the Rhine. Even in Taiwan, proposals are being developed to restore coastal wetlands in Tainan that were reclaimed only 5 years ago.

ELEMENTS OF A SUCCESSFUL WETLAND RESTORATION

21. With about two decades of wetland restoration experience in the U.S. it is worth highlighting some of the key lessons. We now have a perspective on what is needed to achieve successful restoration projects.

(a) Need for competent executive agency. For a wetland restoration project to be successfully implemented requires the leadership of an organisation that has as its primary mission the restoration of natural resources. The authority of such an agency greatly facilitates developing the political and public support, securing funding, facilitating land acquisition, and using the best expertise in biology, hydrology, planning and engineering in the implementation of restoration projects. Fortunately in California we have had the benefit of a model institution to achieve this, the State Coastal Conservancy.

(b) Establishing clear restoration goals. The performance of many early restoration projects became controversial because there was no explicit statement of restoration goals—against which the success of the restoration could be measured.

(c) Understanding the physical system. Most genuine rather than alleged failures in restoration projects stem from a failure to consider properly the hydrologic, hydraulic and geomorphic functioning of the natural system and integrate this into the ecologic design. This partially stems from the lack of research in these areas as compared to the intense interest in wetlands by researchers in the life sciences.

(d) Emphasizing the restoration of the natural evolution of the wetland. The most successful restoration designs have restored the natural physical functioning of the system whose ecology and morphology evolves with little intervention into a valuable wetland. This is to be contrasted with the limited long-term success of highly manipulated or "engineered" systems that require sophisticated management.

(e) Monitoring and learning from experience. The best guide to successful restoration is the practical experience with past projects. Instituting integrated monitoring plans of ecologic and physical criteria, ensuring clear design documentation, and open access to this information is crucial in the future development of projects.

22. Successful wetland restoration typically requires an integration of an understanding of ecology with engineering and hydrology. In the past many engineers have been indifferent or even antagonistic to efforts to protect natural resources from human intervention. This has led to the neglect of the study of the hydrology, hydraulics and geomorphology of wetlands.

TOWARD THE FUTURE

23. The challenges of wetland restoration are a reflection of the new and larger challenges for managing and restoring the global ecosystem in the 21st century, requiring new institutions, the development of new methods, answering new research questions and restructuring educational curricula. To meet this challenge it is perhaps also time to rethink how we have defined our professional disciplines.

24. The civil engineering profession was born in the early 19th century as an instrument of the Industrial Revolution. In 1827, the Institution of Civil Engineers selected a definition of its profession as "The art of directing the great sources of power in nature for the use and convenience of man."

25. In the light of the challenges posed to human society by our own actions in destroying the global environment, and our responsibility for restoring what we have destroyed, it is now time to define our task as "The art of directing the great sources of power in mankind for the benefit of nature."

REFERENCES

1. DUGAN P.J., ed. Wetland conservation: a review of current issues and required action. IUCN, Gland, Switzerland, 1990.
2. MITSCH W.J. and GOSSELING J.G. Wetlands. Van Nostrand Reinhold Company, New York, 1986
3. NATIONAL RESEARCH COUNCIL. Restoration of aquatic ecosystems: science, technology, and public policy. National Academy Press, Washington, D.C., 1992.
4. PURSEGLOVE J. Taming the flood. Oxford University Press, Oxford, 1989.

Lessons learned in wetlands restoration and enhancement

P. GRENELL, Proprietor, Peter Grenell & Associates, San Francisco

SYNOPSIS. Since 1978, the California State Coastal Conservancy has gained experience in protecting, acquiring, restoring, and enhancing wetlands on California's coast and San Francisco Bay. This paper examines the Conservancy's attempt to save and improve California's coastal wetlands, reviews the most important lessons learned, and suggests their potential applicability.

INTRODUCTION

1. The science of wetlands management is an evolving discipline. The California Coastal Conservancy has operated one of the most effective wetland programs in the country (ref. 1). Its experience reflects the gradual increase in understanding of wetlands. It also testifies to how much we still don't know about how they function, the physical and bio-chemical processes and their economic implications affecting wetlands, the impacts of human actions on them, and how we can best protect our remaining wetlands.

THE STATE COASTAL CONSERVANCY'S WETLANDS PROGRAM
Background

2. In 1976, the California Legislature created a unique state agency to address coastal resource management problems that could not be handled by regulatory means. The Coastal Conservancy is a non-regulatory agency that complements the state's two coastal regulatory bodies, the Coastal Commission (usually referred to as the Commission) and the San Francisco Bay Conservation and Development Commission (known as BCDC). The Conservancy's charge was to protect, restore, and enhance coastal resources. From wetlands to waterfronts, from public access to agriculture, the Conservancy uses a variety of land acquisition, planning, technical assistance, and financing tools to design and implement projects along California's 1,100 mile coast and around the nine county San Francisco Bay area. The thrust of the Conservancy's activities has been to resolve land use and resource conflicts through non-adversarial means, without compromising the natural resources (ref. 2). Much of this activity has been directed to wetlands.

SETTING THE SCENE

3. The Conservancy's experience evolved over the past sixteen years through over five hundred projects and expenditure of nearly $200 million. Nearly ninety-five percent of this amount comes from general obligation bonds allocated to the agency by voter-approved ballot propositions. Roughly one-fourth of the agency's funds have been spent on wetlands projects.

The Coastal Conservancy's Wetlands Program: A Synopsis

4. The United States has experienced a significant reduction in wetlands during the past century (ref.3). California has borne the most drastic wetland loss; recent estimates suggest that California's coast retains only fifteen percent of the wetlands existing before European settlement (ref. 4). Coastal and Bay wetlands have been especially vulnerable to diking and filling for agriculture, urban settlement, and port development, as well as impacts of fresh water diversions and agricultural and urban runoff.

5. The Conservancy viewed its wetlands activities from the outset as having the utmost importance. Among its first six projects was the landmark Arcata Marsh project, a restoration of former industrial area on the north coast's Humboldt Bay to salt, brackish, and fresh water marsh (ref. 5). The agency's wetlands program has grown to become the largest of its six major programs (ref. 6). Over $46 million has been spent on nearly 150 projects covering over 24,000 acres.

6. Projects have involved acquisition, restoration, and enhancement of a variety of wetland types. Because the Conservancy's jurisdiction is broader than the Commission's or BCDC's, it can address "upstream" impacts such as erosion, sedimentation, flooding, and damage to anadrymous fish spawning grounds. The wetlands program's scope encompasses watershed and riparian concerns as well as estuaries, lagoons, and other coastal wetlands. Because of the heavily urbanized character of the coastal zone and San Francisco Bay area, the Conservancy has dealt with urban wetland problems having use conflicts involving technical issues of restoration design and also political and economic problems (ref. 7). Much Conservancy effectiveness has been in resolving these conflicts, and project design becomes part of the dilemma's solution.

7. Because of its conflict-resolving mission, the Conservancy has attempted to solve problems stemming from regulatory requirements for mitigation of fish and wildlife habitat losses resulting from coastal development, such as port expansion. The agency has acquired considerable knowledge about problems and limitations of trying to compensate through artificial restoration development impacts on existing wetlands and other aquatic habitats (ref. 8).

EVALUATING THE CONSERVANCY'S WETLAND PROGRAM'S EFFECTIVENESS

Purpose and Methodology

8. In July 1993, the Conservancy published an evaluation of the effectiveness of its wetlands activities, prepared by San Francisco State University's Romberg Tiburon Center for Marine Studies under the direction of Dr. Michael Josselyn (ref. 9). This analysis was designed to provide recommendations to the Conservancy as to how it could more successfully implement future Conservancy-sponsored projects. The information in paragraphs 9-20 comes from Josselyn's report unless otherwise noted. It summarizes the study's chief findings and recommendations, which focus on the technical problems of wetland restoration and enhancement. Subsequent references to the Josselyn report are made in the text.

9. The study evaluated twenty-two Conservancy projects at nineteen sites, from Arcata on the north coast to San Diego, near the Mexican border. Fifteen were coastal and seven were in the San Francisco Bay area. A data base was compiled for each project from planning documents, staff reports, monitoring information, site inspections, and meetings with project sponsors and other concerned parties. Project effectiveness was measured using two qualitative criteria: evaluation of wetland functions and assessment of project success in meeting general criteria of the National Research Council (ref. 10).

10. The study identified eleven chief wetland functions which were considered in the evaluation. These included:

(a) Vegetative diversity and primary productivity
(b) Nutrient transformation/retention
(c) Sediment trapping/erosion control
(d) Flood storage and desynchronization
(e) Groundwater discharge/recharge
(f) Fish habitat
(g) Wildlife habitat
(h) Recreation
(i) Endangered species habitat
(j) Research/education/natural heritage

11. The questions developed by the NRC in 1992 to assess project effectiveness which were used included:

(a) To what extent were project goals and objectives achieved?
(b) To what extent is the restored ecosystem self-sustaining, and what are the maintenance requirements?
(c) If all natural ecosystem functions were not restored, have critical ecosystem functions been restored?

(d) If all natural ecosystem functions were not restored, have critical components been restored?
(e) How long did the project take?
(f) What lessons have been learned from this effort?
(g) Have those lessons been shared with interested parties to maximize the potential for technology transfer?
(h) What was the project's final cost?
(i) What were the ecological, economic, and social benefits of the project?
(j) How cost-effective was the project?
(k) Would another approach to restoration have produced the desirable results at lower cost?

12. Projects evaluated comprised several wetland types, including tidal marshes (8), fresh water marshes (4), riparian wetlands (2), and a bog. Several contained combinations of wetland types (6). Of the twenty-two projects, nine (forty percent) involved land acquisition, seven enhancement or protection of existing conditions, fourteen (about two thirds) restoration of major wetland functions, and fifteen required Coastal Conservancy funding for plan preparation.

13. The most common enhancement measure used was restoration or management of hydrologic conditions on site to create or enhance wildlife benefits. Of the project sample, ten (45 percent) involved water quality improvements, mainly through treatment of secondary treated effluent before discharge into estuarine or coastal waters. Projects also included fisheries improvements and enhancement for rare and endangered species. Fifteen (over two thirds) included provision of public access and nine had educational goals.

14. The evaluation's sample was significant in that the total project value represented was $20 million, or nearly forty-four percent of total Conservancy funds expended on wetlands projects. Average of total project cost for plan preparation was nine percent; most of the funds went for project implementation. The average time per project from planning to completion was 3.7 years. Seventy percent of the projects had monitoring programs; these were generally more recent projects, as the need for monitoring became more evident to program managers.

Findings and Recommendations

15. Findings. The evaluation's overall conclusion was that "...most Conservancy-funded projects were effective in enhancing wetland functions". This relatively high degree of success was considered to result from (a) extensive project planning, (b) frequent interaction of Conservancy staff with other resource and regulatory agencies, and (c) funding of project grantees who are committed to wetland protection.

16. Other major conclusions of the study: Thirteen projects (nearly two thirds) were effective in meeting project goals with respect to either wetland functions or NRC criteria. In most cases, effectiveness was accomplished by achieving most project plan goals. The study noted that frequently one or more goals could not be met, or required remedial action. This was true for eight projects. Only one project was deemed totally ineffective.

17. Although a large number of Conservancy-assisted projects are self-sustaining, many others require constant human intervention for continued satisfactory functioning. This seems to be mainly a result of use of waste water treatment methods in several projects. Other projects require sustained management for catchment and removal of sediment originating in the surrounding watersheds, or to maintain tidal circulation by periodic breaching of sand bars.

18. In any case, the study found that critical wetland functions are being restored by the projects. Many projects also provided effective recreational and educational benefits, but the evaluation suggested that habitat benefits need to be improved in these projects to attain a similar level of effectiveness as these non-habitat benefits.

19. The study concluded that probably the least effective part of the agency's overall wetlands program effort is transference of experience gained from its projects to new efforts. Lack of monitoring has some influence; lack of uniform data collection and reporting in the literature also contributes to this finding according to the study. In addition to the study's findings, it should also be noted that, to a great extent, agency staff energy was directed to project planning and execution, not information dissemination. Budget restrictions have also prevented the Conservancy from developing a satisfactory public information capacity, further limiting its ability to pass on what has been learned.

20. Recommendations. The Josselyn evaluation made the recommendations listed below. As of this writing, the Conservancy has begun to implement them through the continued operation of its program.

(a) Develop a consistent database of Conservancy projects
(b) Provide improved availability of project data
(c) Provide in-service training of Conservancy staff on the scientific aspects of wetland restoration
(d) Provide better tracking of project cost information
(e) Develop standard protocols for project planning and monitoring
(f) Publish results in an accessible manner for use in other efforts
(g) Provide additional funds for contingencies in project implementation and for project maintenance, replacement, and monitoring costs.

SETTING THE SCENE

LESSONS LEARNED

21. Much has been learned from the Conservancy's sixteen years' experience with wetlands projects. Several concerns stand out.

22. <u>Information Management and Dissemination.</u> As the Josselyn study clearly showed, information is a key concern: gathering it, organizing it, making it available to others as a body of experience from which to draw, and keeping it current. To take maximum advantage of the wealth of information potentially available from wetland restoration and enhancement activities, uniform data collection and data base creation are necessary. Projects also need to be monitored so that "mid-course corrections" can be made if needed to ensure project success, and to provide data on project performance regarding whether project goals were achieved, whether other unintended benefits have occurred, on conditions that prevented goal achievement, and project construction and operating costs. Moreover, as the Josselyn study emphasized, transferring this information in readily usable form to others in the field, practitioners and researchers, is vital to enhancing effective wetland protection practice.

23. Pressures to accomplish things quickly, establish a track record of performance and respond to crises, especially if funds are scarce, can result in deferring attention and allocation of resources to information collection and management, as happened with the Conservancy. Over the longer term, sustained and improved performance and success depend upon institutional capacity for learning from experience, and for this such information is essential.

24. <u>The Importance of Planning.</u> The Josselyn evaluation highlighted the importance of planning. The Conservancy's moderate success was due largely to the extensive planning of its projects. This included frequent interaction of staff with counterparts in both regulatory agencies and the federal and state resource agencies that advised the permit authorities.

25. An example of the importance of planning -- and of adequate data about restoration -- is the Conservancy's first, experimental attempt at establishing a mitigation bank in Humboldt Bay in the early 1980s. The Conservancy, at the behest of the Commission and the City of Eureka, agreed to facilitate the city's redevelopment of a waterfront area containing several marsh remnants, each less than one acre in size, by buying a former sawmill site on another piece of bay shoreline and restoring it to marsh habitat of a size greater than the total acreage of the "pocket marsh" remnants but comparable in habitat value. The Conservancy's costs would be reimbursed from fees levied on permit-holders by the Commission as conditions for future development approvals. An evaluation of the mitigation bank site several years after construction observed that the bank was not functioning as anticipated for reasons related to inadequate knowledge of site conditions and the needs of the particular kind of restoration desired at that site (ref. 11). The Conservancy had to implement

remedial measures to correct the problems in order to continue to receive mitigation fee repayments.

26. Later Conservancy projects have included more rigorous pre-implementation planning, as the 1993 Josselyn evaluation study noted. In the process, recent projects have benefitted from advances in knowledge about wetland functioning and restoration. The Conservancy now recognizes that different wetland situations must be treated differently: tidal marsh restoration in Humboldt Bay requires different methods than in the Tijuana Estuary; reliable techniques in Morro Bay may not be applicable without change, if at all, elsewhere on the California coast. A more successful example of planning a mitigation project, for example, is the Conservancy-assisted Huntington Beach Wetlands restoration, (ref. 12).

27. High Level of Human Intervention Needed. As population increases and urbanization spreads, pressures on remaining wetlands will increase. In this respect, California is a bellwether of things to come. Few pristine wetlands remain in this most populous state in the United States of 31 million people. Urban wetlands, which comprise a large proportion of the Conservancy's wetland responsibilities, require a high degree of sustained human involvement to keep the restored wetlands functioning satisfactorily. As the Josselyn evaluation remarked, in most cases, hydrologic conditions play a major role in restoring or generating satisfactory fish and wildlife habitat. Josselyn noted specifically that hydrologic functioning was the most common problem in the project sample. In tidal marsh restorations, tide gates and weirs require regular adjustments to maintain the correct hydrologic regime. Elsewhere, increased sedimentation from actions "upstream" in the watershed often beyond the control of those responsible for the projects caused problems. Also, attempts to revegetate wetlands with native species sometimes have been subject to invasion of exotics, and require continuous attention.

28. Multi-Functional Aspect of Projects Creates Problems and Opportunities. The Coastal Conservancy has been uniquely placed to initiate, respond to, and facilitate efforts to protect, enhance, and restore fish and wildlife habitat while addressing problems of other wetland functions, including flood control, water quality, use of treated waste water, public education, recreation, and agricultural and urban runoff. Urban wetlands are frequently subject to these pressures, as the needs and desires of human settlements cause further impacts on the remnants of California's formerly vast wetland resource. The Josselyn evaluation observed that use of treated waste water was frequently the reason for sustained human intervention in restored wetland functioning; there are other influences, too.

29. The conventional approach to flood control has been a concrete channel to move flood waters as quickly and safely as possible downstream. Often, this single purpose solution had adverse impacts on riparian and downstream wetland habitat, but these concerns were overridden by the need for protection of human life and property. More

recently, attention has been given to designs that address both flood control and habitat protection issues. The Conservancy has been among the most active participants in these efforts in California.

30. As the Josselyn evaluation observed, many Conservancy-assisted projects have had a recreational or educational component. The fact that urban wetlands often provide the only significant open space and natural areas left for metropolitan populations in coastal California creates both another problem and opportunity for those engaged in wetland protection. Wetlands are a natural laboratory for schools, and a source of continuing enjoyment and wonder for urban populations including, but not limited to, amateur ornithologists. Yet human impacts on wetland wildlife, however benevolently motivated, remain a source of concern for those involved in wetland conservation (ref. 13). While access often is not a problem, it has created disturbances that have deterred nesting birds in marshes.

31. Some Conservancy projects have addressed problems of agricultural impacts on wetlands. Agricultural impacts have included pesticide runoff, as with intensive strawberry farming near Elkhorn Slough (ref. 14), cattle grazing (ref. 15), fresh water diversions, and soil erosion. These projects have often involved sensitive and protracted negotiations with land-owners and other interests to obtain mutually satisfactory agreements on project design.

32. The Josselyn study also noted that, overall, habitat benefits must be improved relative to others provided by the Conservancy's projects. This goal is challenging when joined with other aims of multi-purpose projects. Agency staff have found that acquisition and enhancement of a rural wetland not subject to population-driven impacts is comparatively simpler, especially if habitat values are limited to waterfowl habitat. In urban projects, the technical, financial, and political factors are very active in determining the mix of project benefits to be sought. Increasing the proportion of habitat values in these situations requires more education of local participants as to the importance of habitat benefits and development of greater willingness to explore new approaches, more understanding and flexibility of agency staff responsible for flood control or use of treated waste water, and improved information and monitoring regarding restoration and enhancement techniques.

33. <u>Watersheds and Riparian Corridors as the Basis for Wetland Planning.</u> The relative ineffectiveness of parcel-specific wetland protection in stemming the tide of habitat loss has led to awareness that broader intervention is necessary. Regional movements of wildlife and the concept of "wildlife corridors" (ref. 16) have their counterparts in increasing recognition of watershed and riparian-based planning as the most fruitful approach to protecting wetland and riparian habitats. Perception of the inter-relationships of species and habitats and needs for retaining adjacent uplands and buffer areas from human activities, as well as increased understanding of habitat values of open spaces subject to human activity

like agriculture or salt ponds, have reinforced this view. The Conservancy gradually realized that spending public funds on restoring coastal wetlands created little long-term benefit if upstream development, agriculture, and timber harvesting resulted in downstream impacts like sedimentation, flooding, pollution, and destruction of anadrymous species spawning areas that wiped out those public investments.

34. Whether watersheds should be the basis for regional wetland planning is a subject of debate because of issues related to distance and variability of habitat characteristics, especially if mitigation is involved in which compensation is considered on an in-kind basis, as current policy tends to require. In-kind compensation may not be possible within a given watershed, and so caution is needed before concluding that a mitigation issue can automatically be resolved within the same watershed.

35. <u>Mitigation: Possible, But No Panacea, and No Substitute for Unspoiled Habitat.</u> The controversy over mitigation rages: it is either an open door to more wetland destruction or the solution to the bottleneck that prevents badly needed economic growth. Yet slowly some light is being shed on this murky area. As the Conservancy's evaluation study and other commentators have noted, most mitigation projects have not been successful (ref. 17). These efforts have often been poorly planned, inadequately monitored, have generated only minimal habitat value enhancements, have not contributed to regional biodiversity, those required to carry out the mitigation have not had a real interest in a positive outcome, and they have not been adequately regulated. Nonetheless, the Conservancy's experience in mediating mitigation controversies and in implementing a few mitigation projects has generated useful insight into the pre-conditions for successful mitigation, and approaches that might enable mitigation issues to be more readily resolved.

36. For example, in a forthcoming report to the California Legislature on port-related mitigation, the Conservancy concludes that much time, effort, and acrimony could be reduced by cooperative planning among regulatory agencies, permit applicants, and resource agencies; and by regional wetland planning that would identify regional wetland restoration and enhancement needs, priorities and opportunities, and potential mitigation opportunities, thereby providing a better basis for case by case analysis by permit authorities. The study makes several recommendations regarding mitigation banking, alternative approaches to acquisition of mitigation sites, and new approaches to restoration planning for mitigation purposes (ref. 18).

37. <u>Regulation's Influence on Conservancy Wetlands Activities.</u> Coastal laws, regulations, and guidelines have influenced the Coastal Conservancy's wetlands program in certain respects, but not in others. Commission and BCDC administration of, respectively, the Coastal Act and the McAteer-Petris Act, and the United States Army Corps of Engineers' administration of Section 404 of the Clean Water Act, have limited what land-owners can do with wetland property. As one consequence of this regulatory control,

the Conservancy has participated in land purchases to help resolve development permit disputes over conversion of, or potential damage to, wetlands. Second, Conservancy help has been sought to address mitigation issues, especially with respect to prospective port and harbor expansion. The Conservancy has assisted in mitigation project site identification, planning, and negotiations of agreements between permit applicants and regulatory authorities. As a non-regulatory agency, the Conservancy at best can mediate, propose alternative solutions, and provide information to the primary conferees.

38. The Conservancy has thus provided, at least to some extent, a means for helping resolve wetland-related permit issues prior to the adversarial, quasi-judicial permit review. Another effective Conservancy role has not been predicated upon the regulatory framework, and has been more anticipatory in nature: advance land acquisition for resource preservation.

39. Where wetland habitats containing rare and endangered species or other especially important characteristics are threatened by human interventions, regardless of whether a development permit has been applied for, the Conservancy has often acquired such property (or had it purchased with grants to local governments or non-profit groups) for permanent protection (ref. 19). Sometimes purchase is the only way to protect such threatened resources. These advance acquisitions have typically occurred without regulatory intervention.

40. The Conservancy has usually provided additional financial and other aid to restore or enhance these properties once acquired by local governments or eligible non-profit organizations. Restoration and enhancement work is subject to regulatory review, although the Conservancy and its grantees have not typically experienced problems in gaining approvals with one area of exception. Regulatory attention has been given to more innovative projects involving use of treated waste water or combined flood control and habitat enhancement. These projects have generally taken longer from start to finish because of water quality agencies' and permit authorities' concerns about habitat impacts of the treated waste water, and that project designs provide adequate flood protection.

41. In comparison to California's coast, Central Valley wetlands are not protected by a regulatory structure comparable to that on the coast and in San Francisco Bay. The state Department of Fish and Game (DFG) and regional water boards have limited authority regarding stream alterations and water quality issues, and the U. S. Army Corps of Engineers, with its Section 404 authority under the Clean Water Act over filling and disposal of dredged materials in "waters of the United States", is an active regulatory presence as it is on the coast, but on only about one-fourth of all wetlands. DFG's land acquisition arm, the Wildlife Conservation Board, along with non-profit organizations, has acquired much wetland acreage. The acquisitions have been mostly to protect waterfowl habitat; other

habitats and wetland values have received less attention, in contrast to coastal wetlands protection work.

42. <u>The Importance of Project Selection.</u> The Josselyn study concluded that especially when funds are scarce, wetland project selection should focus on those with a high potential for success, which can be implemented with proven restoration methods, and which generate high quality wetland resources (ref. 20). As one response to this recommendation, the Conservancy is bringing together panels of wetland scientists to advise it on, among other subjects, wetland project selection. Tapping available expertise and knowledge in the technical literature is likely to help increase the chances of future success.

43. Such concentration of effort will be even more important as specific projects emerge from watershed, riparian corridor, and other regional wetland planning. Project implementors will need to identify variables which will provide the most beneficial overall impact for the least cost. They must also identify those wetland values subject to the most immediate and severe threats, and potential impacts of those losses on the particular wetland ecosystem if not addressed adequately by the project, regardless of the benefit/cost criterion. "High Success Potential" should be determined with both considerations in mind, however imperfect the available data.

APPLYING THE LESSONS

44. In the Fall of 1993, the Conservancy indicated its intention to implement the Josselyn study's recommendations. One may look forward to greater successes of the agency's wetlands program in the future, as long as funding is available; but what is the potential for broader application of the study's conclusions, and of those others presented in this paper?

45. California's severe economic situation will continue for some time, with fiscal constraints on wetlands protection remaining in place. The information base on wetlands functioning, restoration and enhancement is enlarging, as a result of research efforts in California and around the country. Whether the state has the human and financial resources to make it available on a broad scale and to help expand it is uncertain. Thus far, the only action to be taken in this area by the administration has been to direct a few federal funds to the Conservancy to support publication of a guide for private land-owners to protection and beneficial use of their own wetlands (ref. 21).

46. Nationally, the protection of wetlands has become much more visible as a policy issue over the past decade. The 1988 release of the report of the National Wetlands Policy Forum was a significant, if preliminary step, in the process. Notable was the report's goal statement: In the interim: "To achieve no overall net loss of the nation's remaining wetlands base", and over the long term: "To increase the quantity and quality of the nation's wetlands resource base." (ref. 22) This goal was adopted by then-President Bush, and was re-affirmed by President Clinton.

47. The Association of State Wetland Managers continues to disseminate current information on wetlands protection and restoration. Recent ASWM-convened conferences on mitigation, urban wetlands, and watershed planning cited earlier have focused on the most pressing issues and promising directions and methods; Coastal Conservancy experience has been featured in several of these efforts. Of growing interest have been cooperative wetland projects that involve diverse interest groups in project planning, and thereby tend to avoid or reduce friction over project purposes and design. Cooperative projects are particularly relevant to mitigation. Again, the Coastal Conservancy's experiences with such projects have much to offer as a problem-solving approach in project design (ref. 23).

48. Yet at the national level major constraints exist: jobs, the economy, health services, foreign trade, and other issues crowd out environmental and natural resource concerns, at least as far as significant federal budget allocations are concerned. Further advances in the practice of wetland restoration and enhancement in the United States will continue to be piecemeal for the immediate future, as researchers, consultants, officials, and others involved obtain and use what data and reliable, appropriate methods they can. Meanwhile, humility in the face of nature's wondrous complexity remains in order.

REFERENCES

1. JOSSELYN M. and CHAMBERLAIN S. Wetland Restoration by the Coastal Conservancy: A Status Report. Coast & Ocean, vol. X, no. 1, Winter/Spring, 1994.
2. STATE COASTAL CONSERVANCY. The California Coastal Conservancy. Oakland, 1984; GRENELL P. The Once and Future Experience of the California Coastal Conservancy. Coastal Management, vol. 16, no. 1, 1988, pps. 13-15.
3. THE CONSERVATION FOUNDATION. Issues in Wetlands Protection: Background Papers Prepared for the National Wetlands Policy Forum. The Conservation Foundation, Washington, D.C., 1990, p. 7.
4. JOSSELYN M. et al. Evaluation of Coastal Conservancy Enhancement Projects: 1978-1992. San Francisco State University Romberg Tiburon Center, Tiburon, 1993, p. 1.
5. ARNOLD C. Arcadian Waterfront. California Coast & Ocean, vol. I, no. 2, Spring, 1985.
6. GRENELL P. A New Framework for Integrated Coastal Zone Management. The California Coastal Zone Experience. Coastal Zone 91 Conference-ASCE. American Society of Civil Engineeers. 1991, p. 34.
7. GRENELL P. Coastal Conservancy Roles in Cooperative Urban Wetland Projects. Urban Wetlands and Riparian Habitat. Proceedings of the Association of State Wetland Managers Symposium, June 1988.
8. GRENELL P. The Coastal Conservancy's Emerging Role in Shaping Wetland Mitigation Approaches: Standards and Criteria. Mitigation of

Impacts and Losses. Proceedings, Association of State Wetland Managers, October 1986; GRENELL P. and DENNINGER M. Mitigation Banks and Joint Projects. Symposium on Effective Mitigation: Mitigation Banks and Joint Projects in the Context of Wetland Management Plans. Association of State Wetland Managers, June, 1992.
9. JOSSELYN et al., 1993, Op. Cit.
10. NATIONAL RESEARCH COUNCIL. Restoration of Aquatic Ecosystems: Science, Technology, and Public Policy. National Academy Press, Washington, D.C., 1992.
11. JOSSELYN M. Bracut Wetland Mitigation Bank, Biological Monitoring Final Report. State Coastal Conservancy, Oakland, 1988; RIDDLE E. Wetland Mitigation Banking Case Study Guidebook, prepared for U. S. Army Corps of Engineers Water Resources Support Center. Fort Belvoir, Virginia, July 1992.
12. ELIOT W. and HOLDERMAN R. The Huntington Wetlands: Pickleweed or Parking Lot? California Waterfront Age, vol. X, no. 1, Winter 1988.
13. JOSSELYN M. et al. Public Access and Wetlands: Impacts of Recreational Use, Technical Report No. 9, Romberg Tiburon Center for Marine Studies, San Francisco State University, 1989.
14. MARCUS L. Cultivating Good Neighbors. Coast & Ocean, vol. 7, no. 4, Fall, 1991.
15. HOLDERMAN R. Duck Soup. Coast & Ocean, vol. 7, no. 2, Summer 1991.
16. ARNOLD C. Wildlife Corridors. Coast & Ocean, vol. 6, no. 2, Summer, 1990.
17. JOSSELYN et al., Op. Cit. See also Association of State Wetland Managers. Proceedings of Symposium on Effective Mitigation: Mitigation Banks and Joint Projects in the Context of Wetland Management Plans, June, 1992; Conservation Foundation. Op. Cit., 1990, pps. 185, 187, 219-221; Riddle E. Mitigation Banks: Unmitigated Disaster or Sound Investment? California Waterfront Age, vol. 3, no. 1, Winter, 1987; and Zedler J. Mitigation Problems on the Southern California Coast. Op. Cit.
18. STATE COASTAL CONSERVANCY. Port Mitigation Study. Oakland, 1994 (forthcoming).
19. GRENELL P. Non-Regulatory Approaches to Management of Coastal Resources and Development in San Francisco Bay. Marine Pollution Bulletin, vol. 23, 1991/EMECS '90.
20. JOSSELYN et al., Op. Cit.
21. STATE COASTAL CONSERVANCY. The California Land-Owner's Assistance Guide. Oakland, 1994 (forthcoming).
22. THE CONSERVATION FOUNDATION. Protecting America's Wetlands: An Action Agenda, The Final Report of the National Wetlands Policy Forum. Conservation Foundation, Washington, D.C., 1988, p. 3.
23. GRENELL and DENNINGER. Op. Cit.

Wetland archaeology and wetland management

B. J. COLES, BA, MPhil, FSA, Reader in Prehistoric Archaeology, University of Exeter

SYNOPSIS Wetlands of different types and origins may be managed for a variety of purposes, but how often will their archaeological dimension be considered? This paper examines the significance and ubiquity of wetland archaeology, and considers how the common threats to wetlands affect their archaeological component. Three of the few management schemes designed to protect wetland archaeological sites are briefly described. Management of wetlands for wildlife will potentially have a great effect on archaeology, and this will be primarily beneficial if wetland managers are in a position to recognise and respect the archaeological component of their reserves.

THE ARCHAEOLOGICAL DIMENSION OF WETLANDS

Wetlands archaeology is significant across all continents, but in the paragraphs which follow examples are taken mainly from Britain and Ireland.

1. *Palaeoenvironmental evidence.* In northwestern Europe, wetland deposits have accrued since the last retreat of the ice-sheets, beginning some 10,000 years ago or more. In some cases, there has been a more-or-less continuous history of development, in others the wetland character of a particular location has lasted for a few centuries or a few millennia. For example, a palaeochannel of a river such as the Severn, effectively a wetland only from the time it was isolated from the active river system, might contain a fill dating from the later medieval period to the present. The Solway Mosses, on the other hand, are relics of a complex of peat wetlands which began to develop as the ice-sheets withdrew and which are still active in a few locations today.

2. Almost every wetland contains evidence for environmental conditions during the period of its development. Where the evidence is stratified in chronological sequence, which is often the case, it can be used to reconstruct the pattern of environmental change both within the wetland and for the surrounding area. From the pioneering work in Britain of Godwin (ref.1.) to the diversity of palaeoenvironmental reconstruction today, the science has

gained increasing chronological and spatial precision, and now it may be possible to detect in British wetland deposits the fall-out from an identified Icelandic volcanic eruption (e.g. ref.2) as well as using pollen to determine local vegetational change by the decade if not by the individual year (e.g. ref.3).

3. Because people were living in Britain throughout the post-glacial period (and earlier), and because they influenced the course of environmental change, the environmental record contained in wetland deposits is also a record of human activity, particularly from the time that farming was introduced about 6000 years ago. Decreasing amounts of pollen from tree species and increasing amounts from open-ground plants record the steady encroachment of people on the natural woodland cover of the British Isles, a process that took place on dryland but for which the evidence survives under wetland conditions. In a particular wetland, such as the Somerset Levels, detailed analysis of stratigraphy, pollen and other sources of evidence will provide an outline of the changing mosaic of natural vegetation, pasture and arable which can then be compared with the sequence determined elsewhere, as in Romney Marsh (refs. 4-6).

4. *Cultural evidence.* Direct evidence for human activity comes from structures and artefacts and occasionally surviving parts of humans themselves. Under dryland conditions such evidence survives in a degraded state, and it is frequently wetlands that provide the optimum conditions for preservation. Allied to the palaeoenvironmental record, this gives wetland archaeology a subtlety and complexity which challenges the investigator to a broader and more integrated understanding of the past. Perhaps the most important aspect is the preservation of organic materials, which were commonly used by our predecessors but which do not often survive for long, due to the normal processes of biological decay. In waterlogged anaerobic conditions, bacterial and fungal attack is inhibited and wooden structures such as house foundations or trackways may endure for centuries, along with objects ranging from massive oak canoes to leather sandals.

5. These artefacts of human activity have an intrinsic interest, and they are also valuable to the study of the past because, very often, their context is undisturbed. A canoe lies where it was abandoned, overwhelmed by peat, neither broken up for fire wood nor split into house planks. A lost sandal has not been chewed by dogs or foxes. The upper parts of structures may decay, but those parts that are protected by the wet nature of the matrix that surrounds them, survive and do not suffer further dispersal by ploughing and other subsequent activity. Associations are therefore strong: a smashed pot lies beside a trackway, its contents of hazelnuts spilled through the surrounding reeds.

SETTING THE SCENE

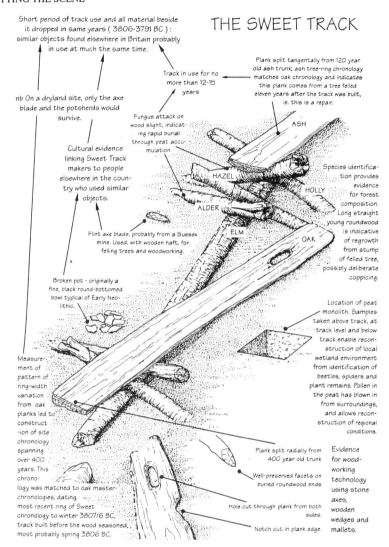

Fig. 1. Combined archaeological and environmental studies on the Sweet Track, Somerset Levels

6. A further important association is that of cultural and palaeoenvironmental evidence (e.g.Fig.1). This comes in two forms: either something such as an oakwood canoe is both cultural (canoe) and environmental (oak tree trunk), or the cultural evidence is surrounded by a matrix which contains contemporary environmental evidence in the form of pollen, macroscopic plant and insect remains, diatoms and so forth. Organic materials also enable the evidence to be dated, normally by radiocarbon dating and increasingly, where oak and some other species are present, by dendrochronology. The

potential is illustrated by the following example from the Somerset Levels, just one short episode from six millennia of interaction between people and wetland.

7. *The Sweet Track.* Early in the history of peat formation in the Levels, when much of the wetland area was dominated by *Phragmites* and the surrounding dryland was covered in primary forest, people came and felled oak and lime, ash, holly and hazel from the forest, alder, willow and poplar from the wetland fringes, and used the wood to make planks, poles and stakes with which they built a raised plank walkway across the reed swamp from a large central island to its southern margin. Their axe blades were made from igneous stone or flint, and using these and wooden wedges and mallets they felled and split trees up to 400 years old and 1 m in diameter. The trackway was used and repaired for about eleven years, after which it was rapidly lost to view in the accumulation of *Phragmites* litter and other peat-forming plant debris. Macroscopic remains of aquatic plants and beetles indicate occasional open pools, and the trackway was substantially reinforced as it crossed one of these, and as they passed here the people who used the track may have watched raft spiders on the hunt, for the remains of *Dolomedes fimbriatus* and *D. plantarius* were preserved in the peat adjacent to and contemporary with the reinforced trackway (ref.7).

8. Dendrochronology has recently dated the trackway to 3807/6 BC (ref.8), i.e. the oak trees used to build it were felled in that winter, and the relative lack of fungal and weevil attack on the wood, coupled with evidence for plank making and track construction before the wood had seasoned, all combine to suggest that the track was made in the year of felling. Fig. 1 illustrates the direct linkage of archaeological and environmental evidence which lies behind the deciphering of what is now known as the Sweet Track; the essential specialist studies are published in the journal *Somerset Levels Papers.*

9. *Distribution of the evidence.* Evidence for the past environment in which people lived, and evidence for how they affected it, is widespread in wetlands and it can be retrieved by taking cores or a small monolith from undisturbed deposits. Cultural evidence has a more restricted spatial and chronological distribution, with considerable apparent variation from one wetland to another. The Somerset Levels are one of the richest British wetlands, in archaeological terms, with many wooden trackways, stray finds of wood, stone and bronze, and settlements of Iron Age date (ref.7). By contrast, Fenns and Whixall Mosses on the Shropshire-Clwyd border have yielded just one bronze axe, one coin and a bog-body of unknown date (ref.9). This may be a reflection of different intensities of exploitation in the past, but it is probably also due to

recent history. In the Somerset Levels, the majority of archaeological finds were made as a result of peat-cutting and monitoring by archaeologists. On Fenns and Whixall, the scale of peat-cutting has been much less and there has not been any long-term, regular archaeological monitoring. The archaeology could be there, but we do not know whether it is or not, and at present techniques of remote sensing in wetlands are in their infancy as far as the detection of archaeological material is concerned (ref.10).

10. When the management of wetlands is being planned, therefore, the archaeological dimension will almost always need to be considered, whether or not there have been detailed archaeological studies of the area. Before turning to this aspect, though, it will be useful to outline the main threats to wetlands archaeology since it is the identification of threats and methods of mitigation that will determine management plans.

THREATS TO WETLAND ARCHAEOLOGY

11. Threats range from 'natural' to 'human', with a very unclear dividing line between the two. A number of the natural threats derive from changing climate, which itself may be due to the greenhouse effect caused at least in part by the industrialisation of human society. Two recently-published surveys from a rapidly developing field of research are *Submerging Coasts* (ref.11 - global) and *Sea-Level Rise and Global Warming* (ref. 12 - southern England); the information which follows has been taken from those publications unless otherwise indicated. Estimates of future sea-level rise in the order of 20cm in the next 35 years or 60cm to 1m in the coming century may not seem much in the working life time of an archaeologist used to the immediate damage caused by urban development or agricultural drainage, but it is generally agreed that there will be increased coastal erosion as wave action reaches further landwards. Archaeological sites at present in the intertidal zone will be exposed and eroded. Sites currently buried just beyond the reach of high tides will be revealed and thereby put at risk. The quality of the potential archaeology in intertidal wetlands is evident from recent work along the coast of south Wales and the Severn Estuary, in the Solent, and along the coast of Essex (e.g. refs. 13-14).

12. Coupled with a rise in sea level, it is predicted that weather patterns could change with increased frequency and severity of storms and increased likelihood of drought. Storms and floods may keep wetland sites nicely waterlogged, but they will also lead to coastal, lake shore and river bank erosion and human attempts at flood control could adversely affect floodplain wetlands. Drought can lead to a rapid fall in water levels (see discussion of

water abstraction, para. 17) with consequent desiccation of bogs and fens, increased tree growth and increased activity of burrowing animals. None of this is good for buried waterlogged archaeology, and organic material will be particularly vulnerable under a régime of increased fluctuation between flood and drought.

13. Acid rain, the result no doubt of human activity, has a wholesale effect beyond the scope of individual human action to control. A survey for English Nature (ref. 15) has shown that, broadly speaking, the north and northwest of Britain receive most acid rain and, mitigated or intensified by local geology, the area most severely affected is north Wales, with south Wales, central southern England and south west and north east Scotland also quite heavily acidified. The implications for waterlogged archaeological sites are as yet uncertain, although cultural material made from wood or iron is thought to be at risk from extreme acidification.

14. Threats resulting from human action vary to a certain extent from one part of Great Britain to another. For example, hydro-electric schemes, which can destroy natural wetland habitats, are more likely in Scotland than in England. The more general threats include peat and gravel extraction, both of which cause severe mechanical damage if not the complete destruction of any sites in their path. Both also require the extensive pumping out of water, which can affect a wide band around the extraction area. Well known examples of archaeological sites threatened in this manner are Market Deeping and Etton, waterlogged sites on the western Fen edge affected by pumping for gravel extraction, and the Sweet Track which was affected by pumping for peat extraction (but now partly protected, see paras 23-24).

15. Some relic wetlands, perceived maybe as marginal or waste land, have been proposed as areas for rubbish tips. Pollution, drainage, sub-surface burning of peat and sundry other unwelcome developments would seem likely consequences, all detrimental to fragile organic evidence, whether it be cultural or environmental. Sewage, from humans and from domestic animals, can cause eutrophication or excessive enrichment of the waters flowing into wetlands. As with acid rain, but probably to a greater extent, this can lead to changes in the natural vegetation cover and possibly to altogether too much vegetation which takes up water on the one hand and roots deeply on the other, thereby causing desiccation and mechanical damage below the surface. The effects of associated chemical changes are as yet uncertain. Agricultural fertilisers, herbicides and pesticides pose problems of eutrophication and pollution similar to those caused by sewage.

16. Possibly the greatest direct threat to wetlands at present is that of

drainage for agriculture, which has caused the loss of so much wetland habitat over recent centuries. As far as archaeology is concerned, it is possibly a more serious threat than that of peat or gravel extraction, because the mechanics of drainage are less obvious than those of peat-cutting or gravel quarrying and no part of the operation allows for the discovery of buried waterlogged evidence. An ancient settlement, concealed 50 cm below a field surface, can lose almost all its significance through the desiccation which follows drainage, though still buried and apparently undisturbed.

17. Water abstraction causes as much damage to wetlands as agricultural drainage has done in recent decades. As population increases and both domestic and industrial demands for water expand, so there is greater groundwater abstraction. Lack of rain over the last few years meant less natural replenishment and increased demand, for crop irrigation for example. The problems have been highlighted by the Water for Wildlife campaign (ref.16), and mitigating action already undertaken by the National Rivers Authority (NRA) demonstrates that the problems are recognised and taken seriously by one government agency at least. In the late 1980s, only 11% of 1310 NRA monitoring stations recorded water flows which were held to approximate natural flows, and it was information such as this which led the NRA to identify river catchments at particular risk where remedial action would be attempted. Low flow in a river means less water in the river-marginal wetlands, and as springs and headwaters dry out, wetlands downstream will suffer. The effects on wetland archaeology are as severe as those caused by agricultural drainage. Excessive groundwater abstraction will also have an adverse effect on isolated waterlogged sites, such as the deposits in the bottom of ancient wells (refs. 17-18). Increased rainfall during the winter months of 1993-94 has provided some immediate relief, but the problems of over-abstraction remain.

18. Development of former wetland areas for urban expansion, light industry, business parks, supermarkets and by-passes has been a common phenomenon of recent years; such development causes extensive damage to any archaeology present, as well as destroying the wetland, because it is inevitably accompanied by drainage, earth-moving and digging trenches for foundations and services. It should be possible to excavate and record any archaeological features in advance of such developments, the work being funded by the developer (PPG 16 applies), but it will rarely if ever be possible to preserve material in situ when a wetland is treated in this way, and conditions may not always allow for the recognition of the archaeology before it is lost.

19. On a smaller scale, wetland management itself can be regarded as a form of development, as can flood defence works. When drainage ditches are blocked to retain water on a reserve, material is often scraped or dug from the immediate environment to build the necessary dams. If a reed-bed is being developed, perhaps to encourage bitterns and other rare forms of wildlife, platforms will be levelled for the reeds to root on, and deep channels or pools excavated to maintain areas of open water within the reeds. Much the same can happen with the creation of a reed filter-bed. Where a reserve is bunded to isolate it form surrounding water because of the poor quality of the latter, or to retain rain-water within the protected area, it is likely that a trench will be dug to insert a vertical impermeable membrane, or a relatively impervious bank will be built around the reserve. All these operations entail disturbance of the ground, which may expose and damage archaeological features as do the operations of commercial developers. In the longer term, management to protect wetlands is likely to benefit any remaining archaeology but, in the run-up to stable wetland conditions, archaeological monitoring of operations, recording of features and perhaps excavation may be necessary.

PROTECTION OF WETLAND ARCHAEOLOGICAL SITES

20. Where archaeological evidence of exceptional quality is known to exist, and where it is at risk because of one or more of the threats outlined above, it may be possible to take some remedial action. So far, attempts to do this have been few, partly because of the expense normally involved and also because the long-term success of such projects is uncertain. The following examples illustrate the scale and nature of attempts to protect waterlogged archaeological sites in situ.

21. *Llangorse, Brecon Beacons, Wales.* The only known crannog in Wales is that of Llangorse where, in the 1st millennium AD, a settlement was built on an island in the lake and its remains now spread off the shores of the island under the lake waters. Excavations by Allan Lane and Mark Redknap have demonstrated the importance of the site (ref. 19), and their work has shown that the remaining underwater deposits are threatened by erosion. The cause appears, in part at least, to stem from the waves created by power boats, the lake being a recreational area within the National Park. Both CADW and the National Museum of Wales were concerned to protect the site, and they were advised that this might be achieved through a relatively simple and low-cost scheme which was put into effect in the autumn of 1993 (Fig.2). In essence, a barrier of staggered hay-bales encircles the area to be protected, allowing water to flow through but breaking the power of the waves. This will encourage siltation behind the barrier, over the archaeological layers, which

SETTING THE SCENE

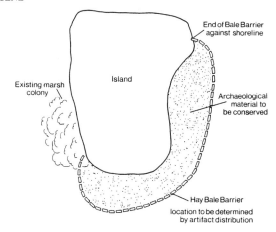

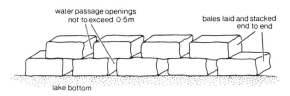

Fig. 2 Llangorse, Brecon Beacons.
Hay-bale protection scheme.

will be further stabilised by shoreline vegetation. There is some concern that plant roots and rhizomes could damage the site - the depth of overburden, water level and plant species are all critical to the control of root penetration - and it may be necessary to delay plant colonisation until sufficient silt has accumulated to protect the site, or at least to ensure that appropriate species are introduced. A further, highly practical step may be taken by the National Park to protect the crannog, namely drafting new bye-laws to ban speed boats from the lake (Redknap, *pers.comm.*).

22. *Corlea, Co. Longford, Ireland* (refs. 20-21). The numerous ancient wooden trackways surviving in Irish peatlands are now recognised and becoming better known thanks to the work of Raftery and the Irish Archaeological Wetland Unit. Most of the trackways have been discovered and then cut away in the course of peat extraction but a substantial stretch of one of the most important tracks has now been preserved, along with its associated raised bog peat. The track is known as Corlea 1, dated to 148 BC, discovered by peat cutting in AD 1984 and excavated by Raftery in subsequent years. Government interest and support, together with assistance from the European Community, has made it possible for the Office of Public Works and Bord na Mona to protect approximately 4 ha of raised bog, with part of the track

displayed in a visitor centre and the remainder undisturbed and waterlogged below the peat (Fig.3). High water levels have been ensured by surrounding the reserve with a buried wall of plastic sheeting and a moat, the details of the scheme having been worked out in consultation with Dutch archaeologists and engineers. Fig. 4 shows how the sheeting has been inserted some way in from the edge of the remaining raised bog, to prevent outward collapse under the pressure of high internal water levels. The folds in the buried sheeting will allow for any vertical expansion in the peat that may take place as it becomes re-wetted, and the bottom edge of the sheeting is dug well into the lower levels of raised bog throughout, to prevent any seepage of water between the two

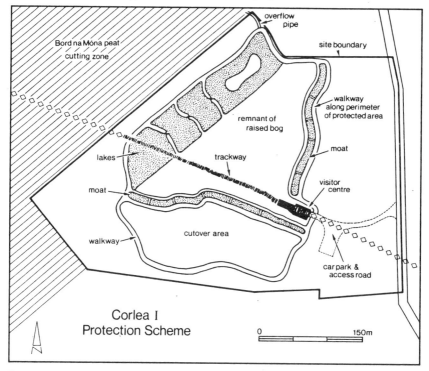

Fig. 3. Corlea, Co. Longford. Protection of Iron Age track in raised bog peat.

distinct levels of peat in the Corlea raised bog. Experience gained in the Netherlands suggested that it was not necessary for the sheeting to penetrate the subsoil, but that the face of the protected area should be further reinforced by dumping of peat against it, which would be consolidated and given a more natural appearance by encouraging the growth of plants appropriate to a raised bog. Clearly, the Corlea scheme pays careful attention to the problems of having a large and heavy body of water retained above the level of the surrounding land. It is one of the most ambitious and comprehensive of efforts

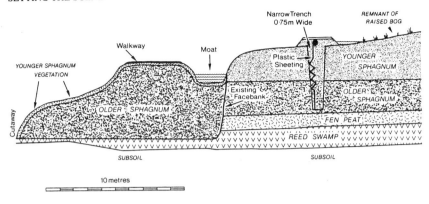

Fig. 4. Corlea Co. Longford. Section to illustrate bunding method.

to protect and display a waterlogged archaeological site, and the raised bog itself will also benefit with likely regeneration and general enhancement of wildlife in the undisturbed areas. The display is of a section of the trackway that was excavated and removed for conservation, which is now returned to the site, under cover, as the main feature of the visitor centre.

23. *The Sweet Track, Somerset Levels, England* (ref. 22). The Neolithic Sweet Track (paras 7-8) was built across a *Phragmites* reed swamp and rapidly engulfed by the build-up of surrounding fen peat. Subsequently, carr woodland and then raised bog conditions led to the accumulation of further layers of peat over the trackway which remained deeply buried, waterlogged and undisturbed until the accelerated peat extraction of recent decades led to its discovery in 1970. Those stretches of the track threatened by peat-cutting were excavated, and in 1980 the line of the remaining lengths of the track was established, and the condition of the wood assessed. Where peat cutting approached its location, drainage threatened the survival of the buried wooden structure; in one place the threatened track passed through an area that was being managed as a nature reserve by the Nature Conservancy Council (NCC, now English Nature), and this too was at risk because of drainage of the surrounding land. For several seasons, the field archaeologists of the Somerset Levels Project monitored water levels within the reserve, and it became apparent that in dry seasons the Sweet Track was not consistently below water throughout the reserve. Prolonged fluctuation would make the waterlogged wood extremely vulnerable to decay, and it was also detrimental to the wildlife interest of the reserve. Thus, in 1983, the archaeological branch of the Department of the Environment (now English Heritage) and NCC together obtained a grant from the National Heritage Memorial Fund which enabled NCC to buy the land of the nature reserve from the owners, Fisons, at a price well below the current

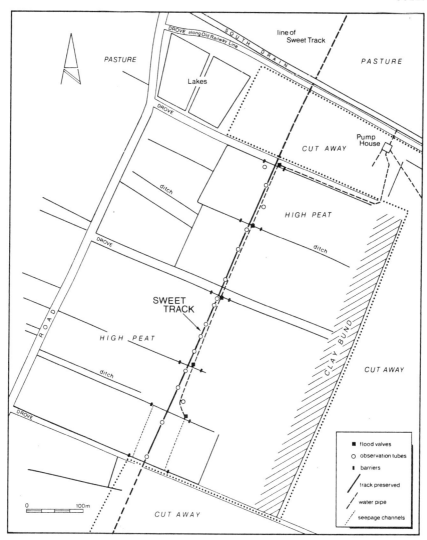

Fig. 5. Shapwick Heath Nature Reserve, Somerset Levels. Combined protection of wetland wildlife and archaeology.

rate for prime peat extraction land. A pumping system was installed to distribute water along the line of the buried track, and a clay bund was built along the edges of the reserve exposed to peat cutting (Fig. 5). There have sometimes been shortages of appropriate water supplies, and sometimes mechanical problems with the pumps, but for ten years both Sweet Track and Shapwick Heath reserve have been kept much wetter than would otherwise have been the case.

24. Now that English Nature are likely to acquire all the surrounding cut-over land from Fisons and will raise water levels over the bulk of the area, it

looks as though it will be easier to keep the reserve wet in future. Possibly there will be no need for pumping, and sufficient reservoirs of rainwater will accumulate to obviate the need of drawing on supplies over-enriched with nutrients. In years to come, the Sweet Track and the Shapwick Heath Nature Reserve may be seen as a pioneering instance of joint archaeological and nature conservation protection, where a relatively high-technology scheme maintained water levels in the short term (10-15 years) whilst conditions developed for the implementation of low-technology long-term conservation.

WETLAND MANAGEMENT FOR WILDLIFE INTERESTS

25. The example of Shapwick Heath and the Sweet Track shows the benefits of co-operation on a site where both the archaeological and the wildlife value is known. But, as indicated above, there are many wetlands where the archaeological component cannot be assessed without putting it at risk: whatever is there is buried, protected, and had better remain so for as long as the integrity of the wetland can be assured. The circumstances can be illustrated by a brief examination of the conditions on West Sedgemoor.

26. *West Sedgemoor.* The moor is a southern tongue of the Somerset Levels and Moors, long and narrow between hills, drained for agriculture but not intensively farmed until quite recently. In the early 1980s conflict arose between those farmers who wanted to increase production from their land through drainage and the then Nature Conservancy Council (NCC) which sought to protect the significant wildlife interest of the moor. By 1986 the Royal Society for the Protection of Birds (RSPB) had acquired a substantial area of the wetland and a fringing wood, but not enough to guarantee control of water levels, and so the process of buying land and shooting rights has continued to the present (1994). Today, 1200 acres or approximately half of West Sedgemoor is owned and controlled by the RSPB and managed in order to improve conditions in the spring and summer for breeding birds such as snipe, redshank, lapwing, curlew and black-tailed godwits. In the winter the flooded or wet grassland provides an internationally important refuge and feeding ground for wildfowl. Varied land usage in the past means that West Sedgemoor today provides a mosaic of different types and qualities of wet grassland, interspersed with land which is not under RSPB control. The management of water levels via ditches and sluices is therefore relatively complex, with some areas deliberately kept wetter than others. Overall, the traditional wetland character of the moor has been enhanced compared with a decade ago, and its inclusion within the Environmentally Sensitive Area of the Somerset Levels and Moors should assist in the long-term improvement and perpetuation of its conservation status.

27. There is peat on West Sedgemoor, but no peat-cutting in recent times and little other opportunity for archaeologists to observe what lies below the surface of the moor. However, the local topography of the long, narrow Moor with 'islands' near the southern end, a freshwater stream, wooded hill slopes, and in the past resource-rich fen vegetation in the wetland opening onto broader fens and the river Parrett to the north, all combine to suggest an area favourable to settlement and exploitation by early peoples. West Sedgemoor is a wetland of high archaeological potential, and current management by the RSPB has greatly improved the chances of survival for any buried, waterlogged organic deposits which survived drainage in the 1970s and 1980s.

CONCLUSIONS

28. There must be many similar wetlands where the archaeological potential is high and where management for wildlife will benefit archaeological interests as well. In order to maximise benefits, managers will need to appreciate the archaeological and palaeoenvironmental dimension of wetlands, and the threats which they themselves pose as 'developers' in the initial stages of their work. Archaeologists will need to forego the possibility of excavation, for the less the deposits are disturbed the better; advances in remote sensing are likely, in the not-too-distant future, to make possible at least a degree of non-invasive investigation of buried waterlogged deposits.

29. One final point. There is much debate over whether or not wetlands can be 'restored' once damaged, and discussion of 'wetland creation'. If a wetland such as West Sedgemoor is drained, sympathetic management can return it to wetland conditions, although probably not to conditions identical to those which existed before drainage. There is no similar way by which wetland archaeology can be restored. Once decayed, the evidence is irretrievable, and it cannot be replaced. Wetland management which takes account of the value and the vulnerability of wetland archaeology across the world (ref. 23) is therefore all the more welcome.

30. *Acknowledgement.* Much of the research for this paper was carried out as part of a survey of wetland management techniques funded by English Heritage, to be published in 1994 (ref. 24).

REFERENCES

1. GODWIN, H. The archives of the peat bogs. University Press, Cambridge, 1981.

2. BARBER K. Peatlands as scientific archives of past biodiversity. Biodiversity and Conservation, 1993, vol. 2, 474-489.

3. BOCQUET, A. et al. A Submerged Neolithic Village. In J.M.COLES and A.J. LAWSON (eds), European Wetlands in Prehistory. Clarendon Press, Oxford, 1987, 33-54.

4. CASELDINE, A.E. A wetland resource: the evidence for environmental exploitation in the Somerset Levels during the prehistoric period. In P. MURPHY and C. FRENCH (eds), The Exploitation of Wetlands. British Archaeological Reports No. 186, 239-266.

5. HOUSLEY, R.A. The environmental context of Glastonbury Lake Village. Somerset Levels Papers, 1988, vol. 14, 63-82.

6. EDDISON, J. and GREEN, C. (eds). Romney Marsh: Evolution, Occupation, Reclamation. Oxford University Committee for Archaeology Monograph No.24, 1988.

7. COLES, B. and J. Sweet Track to Glastonbury. The Somerset Levels in Prehistory. Thames and Hudson, London, 1986.

8. HILLAM, J. et al. Dendrochronology of the British Neolithic. Antiquity, 1990, vol. 64, 210-220.

9. DANIELS, J. Fenn's, Whixall, Bettisfield and Wem Mosses: synopsis plan of management. English Nature and Countryside Council for Wales, 1993 (unpublished).

10. SCHOU JØRGENSEN, M. and SIGURDSON, T. Looking into the wetland - using a new ground-penetrating radar system. NewsWARP, 1991, no.10, 2-5.

11. BIRD, E.C.F. Submerging Coasts: the Effects of a Rising Sea Level on Coastal Environments. Wiley, Chichester, 1993.

12. BRAY, M.J., CARTER, D.J. and HOOKE, J.M. Sea-Level Rise and Global Warming: Scenarios, Physical Impacts and Policies. SCOPAC, University of Portsmouth, 1992.

13. BELL, M. (ed.) Severn Estuary Levels Research Committee: Annual Report 1993.

14. MURPHY, P. and WILKINSON, T. Survey and excavation on the tidal foreshore zone. In J.M. COLES and D. GOODBURN (eds) Wet Site Excavation and Survey. WARP Occasional Paper No.5., 1991, 10-15.

15. RIMES, C. Freshwater Acidification of SSSIs in Great Britain. 1: Overview. English Nature, Peterborough, 1992.

16. HILL, C. and LANGFORD, T. Dying of Thirst: a response to the problem of our vanishing wetlands. RSNC Wildlife Trusts Partnership, Lincoln, 1992.

17. ASHBEE, P., BELL, M. and PROUDFOOT, E. Wilsford Shaft: Excavations 1960-2. English Heritage Archaeological Report No.11, London, 1989.

18. WEINER, J. Bandkeramik wooden well of Erkelenz-Kückhoven. NewsWARP, 1992, no.12, 3-11.

19. REDKNAP, M. Crossing the divide: investigating crannogs. In J.M.COLES and D.M. GOODBURN (eds) Wet Site Excavation and Survey. WARP Occasional Paper No.5, 16-22.

20. RAFTERY, B. Trackways through Time. Headline Publishing, Dublin, 1990.

21. O'DONNELL, T. Conservation of Iron Age roadway and associated peatlands at Corlea, Co. Longford. The Engineers' Journal (Journal of the Institution of Engineers of Ireland), 1993, vol. 46 (3), 44-47.

22. COLES, J.M. The preservation of archaeological sites by environmental intervention. In H. HODGES (ed.) In Situ Archaeological Conservation. Getty Conservation Institute, California, 1986, 32-55.

23. COLES, B. and J. People of the Wetlands. Thames and Hudson, London, 1989.

24. COLES, B.J. Wetland Management: a survey for English Heritage. English Heritage and WARP, WARP Occasional Paper No.9, 1994.

Human pressures on natural wetlands: sustainable use or sustained abuse?

R. S. K. BUISSON, PhD, DIC, BSc, MIWEM, Water Policy Officer, and P. BRADLEY PhD, BSc, Manager, Aquatic Unit, Royal Society for the Protection of Birds

SYNOPSIS   An examination of wetland loss statistics and bird population changes gathered by many authorities indicate that past policies and practices have led to a catastrophic loss of wetlands. This loss has been around half of all those that existed since records have been kept. There have been marked changes in peoples attitudes to the environment and specifically to wetlands over the last 20 to 30 years. These changes have been reflected in international conventions and national laws to protect the environment in general and wetlands in particular. An examination of the most recent indicators of change in the wetland environment - wetland bird population changes - indicates that these policies have not led to a cessation of wetland loss and degradation. It is concluded that world-wide wetlands are still under considerable pressure - our use of them is not sustainable.

INTRODUCTION

> "The continued destruction of wetlands by drainage, exploitation and pollution is the worst act of environmental vandalism being committed on a world wide scale today"
> David Bellamy (ref. 1).

1   This paper, which is presented early in this conference, examines wetland management from a historical perspective. The paper examines the pressures and problems faced by wetlands through a profit and loss account approach, identifying the causes behind the changes and seeking to answer the question: Have we managed our wetlands in a sustainable manner or have we just managed to abuse them?

2   The paper highlights the pressures that wetlands face but does not propose solutions since that is the task of the authors presenting papers later in this conference. The RSPB does have positive solutions to propose and is taking positive action for wetland conservation and management, both in the UK and throughout the world through the partnership of bird conservation bodies - BirdLife International (see for instance ref. 2-4)

3   The definition of wetlands to be used in this paper is taken from the "Ramsar" Convention (the Convention on Wetlands of International Importance especially as Waterfowl Habitat, drafted at Ramsar, Iran in 1971). This is that *"wetlands are areas of marsh, fen, peatland or water, whether natural or artificial, permanent or temporary, with water that is static or flowing, fresh, brackish or salt, including areas of marine water the depth of which at low tide does not exceed six metres."* This is a comprehensive definition, allowing an examination of pressures on wetlands from montane peat bogs to the coastal zone.

THE RATIONALE

4   The commonly used definition of what is sustainable has been developed from the statement of the World Commission on Environment and Development. This defined sustainable development as *"development that meets the needs of the present without compromising the ability of future generations to meet their own needs"*.

5   For sustainable use this has been developed into a description of a particular way of using natural resources - one that uses them within their capacity for renewal (IUCN/UNEP/WWF 1991). This concept of the stewardship of natural resources is of course not new but it is one that has rarely been followed. Before the adoption of the term sustainable use in international conventions and agreements, "wise use" was used. In particular this has been the goal of wetland conservation and management as promoted by the Ramsar Convention - an international convention specifically aimed at wetland conservation. A definition of wise use has been adopted by the contracting parties to the Convention. This is that *"the wise use of wetlands is their sustainable utilisation for the benefit of humankind in a way compatible with the maintenance of the natural properties of the ecosystem"*. Sustainable utilisation has been defined as *"human use of a wetland so that it may yield the greatest continuous benefit to present generations while maintaining its potential to meet the needs and aspirations of future generations"*. Natural properties of the ecosystem have been defined as *"those physical, biological or chemical components, such as soil, water, plants, animals and nutrients, and the interactions between them"*.

6   Taking this rationale it is possible to examine for signs of sustainable use and in particular whether or not natural properties of the ecosystem have been maintained.

THE METHOD

7   This paper takes as its hypothesis: If we had been managing wetlands in a sustainable or wise manner then we

would have received from our forebears an equal set of assets as they inherited and that we would be passing onto to the next generation an equal set of assets. This can be examined through wetland loss statistics and associated changes in wetland animal populations.

8   It would be possible to test for changes in pressures upon wetlands by examining the changing statements of Governments and statutory bodies upon their policies for the protection and management of wetlands. Regrettably, they all have their own political credibility and powerbases to defend and cannot be considered an independent source from which to measure changes in pressures upon wetlands.

9   Two approaches are adopted, based upon limitations in the availability of reliable information. The first is an examination of the change over time in the absolute area of wetlands in the world. This gives a basic measure in the change in the quantity of wetlands but gives no indication of changes in wetland quality. The degradation, as opposed to loss of wetlands, requires more subtle indicators. The indicators that have been chosen are wetland bird population statistics. Bird populations respond not just to the area of wetland available to them but also the quality - they can indicate wetlands that are degraded through factors as varied as pollution and recreational use. Unfortunately with the exception of total losses of populations, that is extinctions, reliable bird population statistics are only available for recent decades. This limits the use of this combined indicator of wetland loss and degradation to the 1960's onwards. This is the period when the concepts of sustainable or wise use were elaborated and in theory, enacted.

THE PROFIT AND LOSS ACCOUNT - OUR INHERITANCE

10   A summary of wetland loss statistics extracted from regional summaries prepared by other authors who have totalled the losses going back several centuries is given in Table 1. This shows wide variations in the percentage and rate of loss of wetlands between different countries and regions. In all cases the losses are substantial. In the more highly settled and technologically developed regions such as California's central valley the losses can be virtually absolute.

11   In this historical context, the available information upon bird populations is rather crude, with the most certain information relating to extinctions. Ninety seven species have become extinct since 1600 (ref. 19). The majority (90%) of these species were confined to islands. Approximately one third of the species that have gone extinct were wetland birds. Other bird species have declined considerably and come close to extinction or exist in such small numbers that they are vulnerable to extinction. These species are listed in "Red

Table 1: A summary of wetland loss statistics

Country	Locality	Wetland type	Timescale start	finish	% loss	Rate of loss Ha/annum	Reference
USA	Lower 48 states	All wetlands	1780	1980	53%	188000	6
		All wetlands	1950	1970		185000	7
	Lower 48 states	Freshwater	1974	1983	3%	116000	8
		Saltmarsh	1974	1983	2%	3150	8
	Florida	Everglade wetlands	1900	1990	50%		1
	California	Central valley wetlands	1850	1990	99%	113000	1
			1939	1985		435	1
Canada	Lake Ontario	shoreline wetlands	1850	1990	90%		1
	Fraser River delta		1976	1982	28%		1
	Atlantic coast	Saltmarsh	1800	1990	66%		9
Colombia	Cauca River	Floodplain wetlands	1950	1990	88%	336	10
	Magdalena R Delta	Mangroves	1970	1987	80%	3450	10
France	Brittany	coastal wetlands	1960	1990	40%		1
	Camargue	Wetlands	1942	1984		100	11
Greece		All wetlands	1920	1990	60%		11
	Macedonia	Marshland	1930	1990	94%	1550	11
Italy		All wetlands	1891	1991	77%	10000	11
Tunisia		All wetlands	1881	1991	15%	190	11

Table 1 (contd): A summary of wetland loss statistics

Country	Locality	Wetland type	Timescale start	finish	% loss	Rate of loss	Reference
Denmark		Freshwater wetlands	1840	1990	75%		1
Holland		All wetlands	1950	1985	55%		12
Germany		Bogs and fens	1981	1985	9%	2500	1
Belgium		All wetlands	1960	1990		1000	1
	Flanders	Floodplain pastures	1960	1990	90%	1050	1
United Kingdom		Raised bog	1870	1990	94%		13
England	Lancashire	Peat "mosses"	1948	1975	95%		14
England	Thames Estuary	Coastal pastures	1930	1980	64%		15
Northern Ireland	Belfast Lough	Intertidal Flats	1750	1990	85%		16
Kazakhstan	Aral Sea	Open water	1950	1990	66%	106000	1
Cambodia	Mekong river	Floodplain forest	1970	1990	20%	6800	1
Vietnam	Mekong delta	Mangrove & Melaleuca	1960	1975	54%	10100	1
Thailand		Mangroves	1918	1987	78%		17
			1961	1979	21%	4330	1
Japan		Intertidal flats	1979	1985	7%	679	18
New Zealand		All wetlands	1840	1990	90%		1
		Freshwater wetlands	1979	1983	14%		1
Australia	Victoria	All wetlands	1830	1990	33%		1

Data Books" for individual continents. Some 16% of these species vulnerable to extinction are wetland birds (ref. 19).

THE PROFIT AND LOSS ACCOUNT - WHAT WILL WE HAND ON?

12 It should come as no surprise to face the fact that in the past man's activities have resulted in considerable wetland losses and the extinction of a significant number of wetland birds. Have we learnt the lesson that our activities are damaging the environment in recent years?

13 If the environmentally aware practices which we hear so much about, and will probably hear more of in future sessions of this conference, are put into practice then we should not expect to observe significant signs of wetland loss and degradation. Using the detailed information that is available upon bird population changes for recent years, it is clear that populations of many wetland bird species are declining. In Europe (for this study the Azores to the Urals and Greenland to Turkey) an analysis has been carried out of changes in breeding and wintering bird numbers from 1970 to 1990 (ref. 20). The purpose was to define which species were of concern and hence in need of special conservation measures because their populations were globally threatened, concentrated in Europe or declining significantly. This study has revealed that of the approximately 450 species regularly occurring in Europe, 278 are of conservation concern. One third of these are wetland species. Those wetland species which are globally threatened (ref. 21) and those which have an unfavourable conservation status in Europe because their populations have declined by at least 20% in a significant area (greater than one third) of Europe in the period 1970-1990 are listed in Table 2. Some 73 species, nearly 40% of the total species falling into these categories, are wetland birds. The globally threatened species range from a large fish eating predator - the Dalmatian pelican, to a small brown songbird of eastern European - the aquatic warbler.

14 It could be conjectured that the figures for Europe as a whole conceal the fact that in such an environmentally aware country as the UK, wetlands and their associated birds fare a little better. An examination of the changes in the UK breeding bird populations reveal that this is not the case. A detailed survey of the breeding birds in Britain and Ireland was carried out between 1968 and 1972 and repeated in 1988 to 1991 (ref. 22). Some of the most dramatic declines in breeding bird populations have occurred amongst the breeding waders of the lowland wet grasslands of England and Wales. These birds are dependent upon floodplain pastures grazed at low density of cattle or cut for hay late in the season. Examples of significant declines include snipe which have disappeared from 60% of the areas and redshank from 40% of the areas where they occurred at the time of the first survey.

Table 2. Wetland bird species which are globally threatened (ref. 21) and those which have an unfavourable conservation status in Europe because their populations have declined by at least 20% in a significant area of Europe in the period 1970-1990 (ref. 20)

GLOBALLY THREATENED WETLAND SPECIES IN EUROPE	
Dalmatian pelican	White-headed duck
Lesser white-fronted goose	Sociable plover
Red-breasted goose	Slender-billed curlew
Ferruginous duck	Audouin's gull
Steller's eider	Aquatic warbler

WETLAND SPECIES WITH AN UNFAVOURABLE CONSERVATION STATUS IN EUROPE	
Red-throated diver	Avocet
Black-throated diver	Collared pratincole
Pygmy cormorant	Black-winged pratincole
White pelican	Kentish plover
Bittern	Greater sand plover
Little bittern	Caspian plover
Night heron	Spur-winged plover
Purple heron	Knot
White stork	Dunlin
Black stork	Broad-billed sandpiper
Glossy ibis	Jack snipe
Spoonbill	Great snipe
Greater flamingo	Black-tailed godwit
Barnacle goose	Bar-tailed godwit
Brent goose	Curlew
Ruddy shelduck	Redshank
Gadwall	Wood sandpiper
Pintail	Little gull
Garganey	Common gull
Red-crested pochard	Gull-billed tern
Scaup	Caspian tern
Harlequin duck	Sandwich tern
Velvet scoter	Roseate tern
Barrow's goldeneye	Little tern
Smew	Whiskered tern
White-tailed eagle	Black tern
Osprey	Kingfisher
Baillon's crake	Sand martin
Purple gallinule	
Crested coot	
Crane	

15 For the wintering waders and wildfowl of our coasts and large inland waterbodies such as lochs and reservoirs, the indicators give less worrying signs of recent management of these habitats. Indices of wintering wildfowl and waders for the period 1973 to 1993 show stable or increasing wintering populations for the great majority of wetland species (ref. 23).

THE PRESSURES PUT ON WETLANDS - THE CAUSES BEHIND THE NET LOSS

16 There is clear evidence of a significant historical loss of wetlands. Information is presented above which reveals that loss and degradation, as indicated by declining wetland bird populations, is occurring up to this day. What are the causes of these losses - what are the pressures on wetlands?

17 The direct causes of wetland loss, that is the land use that takes place after wetland destruction, are many and varied. Throughout the world though the single greatest cause of loss has been the conversion of wetlands to more intensive agricultural land. Urban growth and industrial expansion have been significant at some sites, particularly in the more developed world. Numerous other factors also account for wetland loss including recreational development, river regulation for hydropower, and peat harvesting. These causes of wetland loss and the losses to natural processes and extreme events listed by the type of wetland affected are summarised in Table 3. A consideration of outright loss does not tell the whole story - pollution and incompatible land uses can destroy the value of a wetland for wildlife without destroying the wetness of a wetland. This is the value of studying indicators of wetland quality such as wetland bird populations.

18 Behind the cause of wetland loss lies a decision by an individual, community, institution or business to seek for itself a greater benefit from the land occupied by, or water present in, the wetland. Either consciously or subconsciously the decision may have been made based on the attitude that the wetland is a wasteland, only fit for putting to a more "productive" use. This view prevails because the individuals involved do not appreciate the value of wetlands either to themselves, others in the community or to wildlife. These values can either be quantified in monetary terms and or in some cases have intangible values which we should recognise for their own sake. Such wetland values range from the recharge of groundwaters and acting as a flood storage area to being a source of harvestable, renewable resources and a habitat for wildlife. A summary of wetland values is given in Table 4.

Table 3. The causes of wetland loss (adapted from ref. 24)

CAUSE	WETLAND TYPE				
	Estuary	Open coast	Flood plain	Fresh marsh	Peat land
Human actions - Direct					
Drainage	***	***	***	***	***
Dredging	***	-	-	*	-
Built development	***	***	***	***	-
Aquaculture	***	-	-	-	-
Pollution	***	***	***	***	-
Mineral extraction	*	*	*	-	***
Groundwater abstraction	-	-	*	***	-
Human actions - Indirect					
Sediment starvation	***	***	***	***	-
Hydrological alteration	***	***	***	***	-
Subsidence due to mining	***	*	***	***	-
Natural causes					
Subsidence	*	*	-	-	*
Sea level rise	***	***	-	-	-
Drought	***	***	***	***	*
Hurricanes/storms	***	***	-	-	*
Erosion	***	***	*	-	*
Biotic effects	-	-	***	***	-

Key : *** : Common and important cause of wetland degradation and loss
  * : Present but not a major cause of loss
  - : Absent or exceptional

19 It is unfortunate that the aid to decision making which is meant to quantify such values in order to examine if their loss is outweighed by some other community benefit - cost benefit analysis - has in practice rarely been used to identify the benefits that wetlands bring. There are cases where a comprehensive cost benefit analysis has been carried out and the claimed benefits of the "development" (which results in the destruction of a wetland) has been outweighed by the benefits of conserving the wetland (eg ref. 3).

Table 4. The value of wetlands

VALUES THAT CAN BE QUANTIFIED IN MONETARY TERMS
water recharge natural shore and bank protection flood storage sediment trap carbon sink pollutant sink and purifier harvestable, renewable resources peat source recreation site
VALUES THAT ARE DIFFICULT OR IMPOSSIBLE TO QUANTIFY IN MONETARY TERMS
wildlife habitat landscape

CONCLUSION

20 The examination of wetland loss statistics and bird population changes indicate that past policies and practices have led to a catastrophic loss of wetlands. This loss has been around half of all those that existed since records have been kept. The marked changes in peoples attitudes to the environment and specifically to wetlands over the last 20 to 30 years have not been reflected on the ground. One indicator, the changes in wetland bird populations, reveal deficiencies in the effectiveness of Government policies and institutional practices to protect and manage wetlands in a nation as environmentally advanced as the UK. The UK Government of course does not agree with this statement. In its submission to the Commission on Sustainable Development (ref. 25) it stated "*the United Kingdom considers that these arrangements enable it to achieve the environmentally sustainable management of water resources, including the protection of aquatic ecosystems and freshwater living resources*".

21 Auditing the state of wetlands in the future should be considerably eased by the large libraries of aerial photographs and satellite images that are being compiled around the world. These could be used to provide definitive measures of wetland loss even in countries where access on the ground is prevented for political reasons. The databases upon bird populations co-ordinated by BirdLife International and the International Waterfowl and Wetlands Research Bureau will provide an additional measure of the state of the world's wetlands - a measure of wetland health.

22 World-wide wetlands are still under considerable pressure - at present, our use of them is not sustainable. What will the future hold for our children?

REFERENCES

1   DUGAN P.J. Wetlands in Danger. Mitchell Beazely, London, UK, 1993.

2   GUBBAY S. Recommendations for a European Union Coastal Strategy. BirdLife International, Cambridge, UK, 1994.

3   HOLLIS G.E., ADAMS W.M. and AMINU-KANO M. (Eds). The Hadejia-Nguru Wetlands: Environment, Economy and Sustainable Development of a Sahelian Floodplain Wetland. IUCN, Gland, Switzerland and Cambridge, UK, 1993.

4   ROYAL SOCIETY FOR THE PROTECTION OF BIRDS. Wet Grasslands - What Future? RSPB, Sandy, UK, 1993.

5   IUCN/UNEP/WWF. Caring for the Earth. A Strategy for Sustainable Living. IUCN, Gland, Switzerland, 1991.

6   DAHL T.E. Wetlands Losses in the United States 1780's to 1980's. US Dept of the Interior, Fish and Wildlife Service, USA, 1990.

7   TINER R.W. Wetlands of the United States: Current Status and Recent Trends. US Dept of the Interior, Fish and Wildlife Service, Washington, 1984.

8   DAHL T.E. and JOHNSON C.E. Status and Trends in Wetlands in the Conterminous United States, mid 1970's to mid 1980's. US Dept of the Interior, Fish and Wildlife Service, USA, 1991.

9   ENVIRONMENT CANADA. The Federal Policy on Wetland Conservation. Environment Canada, Lands Directorate, Canada, 1991.

10 NARANJO L.G. Ecological Change in Colombian Wetlands. In: Waterfowl and Wetland Conservation in the 1990's - a Global Perspective. Proc of an IWRB Symposium, St Petersburg Beach, Florida, USA, 12-19 November 1992. IWRB, Slimbridge, UK, 1993.

11. ANON. A Strategy to Stop and Reverse Wetland Loss and Degradation in the Mediterranean Basin. IWRB and Regione Friuli-Venezia Giulia, Trieste, Italy, 1992.

12 JONES T.A. and HUGHES J.M.R. Wetland Inventories and Wetland Loss Studies - a European Perspective. In: Waterfowl and Wetland Conservation in the 1990's - a Global Perspective. Proc of an IWRB Symposium, St Petersburg Beach, Florida, USA, 12-19 November 1992. IWRB, Slimbridge, UK, 1993.

13   ROYAL SOCIETY FOR THE PROTECTION OF BIRDS and PLANTLIFE. Out of the Mire: A Future for Lowland Peat Bogs. RSPB, Sandy, UK, 1993.

14   NATURE CONSERVANCY COUNCIL. Nature Conservation in Great Britain. NCC Peterborough, UK, 1984.

15   EKINS R. Changes in the Extent of Grazing Marshes in the Greater Thames Estuary. RSPB, Sandy, UK, 1990.

16   WELLS J. Waste Disposal and Conservation in Northern Ireland. RSPB, Belfast, UK, 1988.

17   SCOTT D.A. Wetland Inventories and the assessment of Wetland Loss: A Global Overview. In: Waterfowl and Wetland Conservation in the 1990's - a Global Perspective. Proc of an IWRB Symposium, St Petersburg Beach, Florida, USA, 12-19 November 1992. IWRB, Slimbridge, UK, 1993.

18   TAKAHASHI S. A National Survey of Surface Water in Japan. In: Waterfowl and Wetland Conservation in the 1990's - a Global Perspective. Proc of an IWRB Symposium, St Petersburg Beach, Florida, USA, 12-19 November 1992. IWRB, Slimbridge, UK, 1993.

19   MOUNTFORD G. Rare Birds of the World. Collins, London, UK, 1988.

20   TUCKER G.M., HEATH M.F., TOMIALOJC L. and GRIMMETT R.F.A. Birds in Europe: Their Conservation Status. BirdLife International, Cambridge, UK, (in press).

21   COLLAR N.J., CROSBY M.J. and STATTERSFIELD A.J. Birds to Watch 2. BirdLife International, Cambridge, UK, (in press).

22   GIBBONS D.W., REID J.B. and CHAPMAN R. (Eds). The New Atlas of Breeding Birds in Britain and Ireland. Poyser, London, UK, 1993.

23   WATERS R.J. and CRANSWICK P.A. The Wetland Bird Survey 1992-93: Wildfowl and Wader Counts. BTO/WWT/RSPB/JNCC, Slimbridge, UK, 1993.

24   DUGAN P.J. Wetland Conservation: A Review of Current Issues and Regional Action. IUCN, Gland, Switzerland, 1990.

25   U.K. GOVERNMENT. UK Report to the Commission on Sustainable Development. HMSO, London, UK, 1994.

Pressures on wetlands

Dr J. GARDINER, Technical Planning Manager, National Rivers Authority, Thames Region

Introduction: Definition of Wetlands
According to the Convention on Wetlands of International Importance, wetlands are defined as:

> 'areas of marsh, fen, peatland or water, whether natural or artificial, permanent or temporary, with water that is static or flowing, fresh, brackish or salt, including areas of marine waters, the depth of which at low tide does not exceed six metres (and) riparian and coastal zones adjacent to the wetlands or islands or bodies of marine water deeper than six metres at low tide lying within' Ramsar Bureau (1989)

This definition is not easily applied in the UK, and even less so for the National Rivers Authority (NRA). With the caveat that the NRA's definition is developing and has yet to be tested and confirmed, it is currently held that wetlands are:

> 'areas of waterlogged and periodically inundated land which support a distinctive and characteristic wetland vegetative type' (marsh grass, rushes, sedges etc).

This considerable difference can create problems in classification and translation of international documentation. However, translation is not too great a problem for the information in Table 1, which relates the roles and elements of wetlands to their function, importance to humankind and unwise use.

The functions and importance to humankind recorded in Table 1 illustrate the concept of ecosystem carrying capacity which has been identified in 'Caring for the Earth' as vital for sustainable development, which is defined as:

> 'improving the quality of human life while living within the carrying capacity of supporting ecosystems'
> (UNEP/IUCN/WWF, 1991)

TABLE 1: Wetland Functions and their human utilisation (Hollis et al, 1987)

ROLE	ELEMENTS	FUNCTION	IMPORTANCE TO HUMANKIND	UNWISE USE
Store/sink	Rare, threatened or endangered plant and animal species and communities.	Genetic diversity Recolonisation source	Gene pool Science/education Tourism Recreation Heritage	Excessive or uncontrolled harvest Damage removal or pollution
	Representative plant and/or animal communities	Ecological diversity Habitat maintenance	Gene pool Science/education Tourism Recreation Heritage	Excessive or uncontrolled harvest Damage removal or pollution
	Peat	Nutrient, contaminant and energy store Habitat support Water storage	Fuel Paleo-environmental data Horticultural use heritage Medicinal products	Drainage Harvest faster than accumulation Destruction
	Human habitation sites	Archaeological remains	Heritage/cultural Scientific Recreation	Destruction Lowering the water table
Pathway	Terrestrial nutrients, water and detritus	Food chain support Habitat support	Food production Water supply Waste disposal	Interruption or abnormal change of flows Pollution
	Tidal exchanges of water detritus and nutrients	Food chain support Habitat support Nursery for aquatic organisms	Fish, shellfish and other food production Waste disposal	Pollution Barriers to flow Dredge and fill
	Animal populations	Support for migratory species including fish	Harvest Recreation Science	Overexploitation Interruption of migration routes Obstruction Habitat degradation
	Lakes and rivers	Waterways	Navigation	Obstruction Reduced flows and levels
Buffer	Water bodies, vegetation, soils and depressions	Flood attenuation	Reduced damage to property and crops	Filling and reduction of storage capacity

TABLE 1 (continued): Wetland Functions and their human utilisation (Hollis et al, 1987)

ROLE	ELEMENTS	FUNCTION	IMPORTANCE TO HUMANKIND	UNWISE USE
Buffer	Water bodies, vegetation, soils and depressions	Detention and retention of nutrients	Food production Improved water quality	Removal of vegetation Drainage and flood protection
	Water bodies, vegetation, soils and depressions	Groundwater recharge and discharge	Water supply Habitat maintenance Effluent dilution River fisheries Navigation	Reduction of recharge Overpumping Pollution
	Water bodies and peat	Local and global climate stabilisation	Equable climate for agriculture and people	Desiccation
	Water bodies	Large volume Large area	Cooling water	Drainage Filling Thermal pollution
Producer	Production of plants	Food, materials and habitat for migratory species and grazing animals	Harvest of timber, thatch, fuel and food Science Recreation	Overgrazing Over-exploitation Drainage Excess change to dry land or other agricultural uses
	Animal production	Fish, shellfish, grazing and fur-bearing animals	Harvest and farming	Over-exploitation Excess change Habitat degradation Drainage Desiccation
	Organic matter	Methane production Nutrient cycling	Fuel Plant growth	Drainage Desiccation
Sink	Lakes, deltas floodplains	Sediment deposition and detention	Raised soil fertility Clean downstream channels Improved water quality downstream	Channelisation Excess reduction of sediment throughout
	Lakes, swamps and marshes	Bio-chemical self-purification Nutrient accumulation	Natural filter for contaminants Treatment of organic wastes, pathogens and effluents	Destruction of the ecosystem Over-loading of the system

It is clear from Table 1 that wetlands as individual ecosystems are vital not only to wildlife and people at local and catchment levels, but also as a significant part of the biological contribution to global homeostasis, or self-regulating equilibrium. As these many functions have become better understood, concern has grown over the effectiveness of measures to secure protection for the remaining wetland areas (Ramsar Bureau definition), reduced from 6% to perhaps 3% of the earth's surface through dredging, filling, draining and ditching, not including areas degraded by pollution (Meadows, 1992).

This concern is not without foundation, given the rate at which losses of both freshwater and coastal wetlands have occurred globally; according to Meadows, wetlands are probably more endangered than forests. These losses have many causes, and this paper explores some of the causes of physical damage or loss, acknowledging that to consider causes of pollution would extend the remit so much as to result in superficial coverage only.

The question of what effect further diminution of the wetland resource will have presents a great challenge to investigators, but how to reduce the pressures on the remaining areas is a challenge of at least equal complexity, since it must deal with institutional factors and the basic 'human condition'. The attitudes of people towards wetlands has always varied according to their culture (Holdgate, 1990), and a prescription which may prove successful in one area may need modifying elsewhere.

One of the major texts to have been produced in recent years emphasising the essential relationship between people and wetlands was the proceedings of an international conference at Leiden, The Netherlands (Marchand and Udo de Haes, 1989). The many case studies underline the contents of this paper, and bring out the significance of public involvement, including work initiated by the author (Fordham et al, 1989), in the search for an approach in accord with the principles of sustainable development. It is 'vitally important that the public understands and supports the adoption of a more sustainable approach to development.' (Scottish Natural Heritage, 1993).

Following this brief introduction to the problem, this paper attempts to summarise the many pressures on wetlands together with their primary, secondary and tertiary causes, the likely effects of continuing loss, the question of recovery and some approaches to sustainable management and mitigation which could accompany more general progress towards making sustainable development operational.

Primary Causes of Wetland Loss

A primary cause of the physical loss of wetlands can be traced to treating the environment in general (and water in particular) as a 'free good', an economist's way of saying that the environment is treated as having no value, intrinsic or otherwise, until it is used to produce goods which are tradeable. This has led to massive exploitation of a finite resource, only part of which can be described as renewable. The confidence underlying such exploitation stems from the seventeenth century Cartesian idea of 'Mind over Matter', translated by Francis Bacon into the infinitely more dangerous idea of 'Man over Nature', achievable through scientific investigation (Brennan, 1994).

The rational reasoning of scientific reductionism, analysing the interactions between elements of a system in the belief that the whole is the sum of the parts (Jones, 1987), led to the fragmentation of system management. The wonders of medicine, science and engineering introduced by the industrial age led to a widespread trust in the 'technological fix' which has only relatively recently been severely questioned in terms of its ecological, social and even economic 'sustainability'.

This trust also reinforced an earlier primary cause of wetland loss. Many authors have pointed to an underlying psychological reason for draining wetlands while ignoring the lack of recovery or rehabilitation (any adverse impact could be technically overcome). Mankind is held to have a deep-seated and abiding fear and loathing of mires, marsh or bog -the very words sounding more menacing than 'wetlands' - because of the belief that they were sources of evil and disease.

Even if this primeval fear no longer directly sponsors wetland loss, the belief in technology and traditional economic theory certainly does. Formidable efforts have been mounted down the ages to drain wetlands, largely for agriculture in response to increases in population, a further primary cause. Perhaps the best known early example of large-scale works in England was the drainage of the fens in the seventeenth century by Vermuyden (Purseglove, 1988, p55), while echoes of the heroic drainage activities which helped the UK's 'Dig for Victory' campaign during World War II to achieve near self-sufficiency were grant-aided by government until 1985.

Elsewhere this pressure continues, often for export earnings, and is likely to do so while economics prescribes value to short-lived tradeable commodities but finds difficulty with the valuation of natural ecosystems on which the quality of human life, if not life itself, may ultimately depend. The greatest disaster along these lines is probably the destruction of the Aral Sea; it is hard to believe this was actually 'planned'.

Global Examples - Secondary and Tertiary Causes

It is instructive to compare the list of secondary causes of wetland mismanagement and loss, (particularly in Greece), in Table 2 (Pergantis, 1991), with the tertiary causes summarised in Table 3 as physical threats, at least one of which has been identified at all the important Greek sites (Valaoras, 1991). Of the secondary causes, 15 of the 39 listed would be

TABLE 2: SECONDARY CAUSES OF WETLAND MISMANAGEMENT

1. Unfortunate consequences of Greek history
2. Insecurity of maintaining and improving the standard of living
3. Economic demand not being fulfilled by the available means
4. Political instability bringing personal insecurity
5. Poor primary level education
6. Conflicting interests and manipulation of information and "power-profit" oriented decision-making
7. Lack of responsibility at most levels
8. Lack of normal democratic processes in most levels of public life
9. Widespread corruption
10. Lack of proper public participation in decision-making
11. Lack of real power for local/regional authorities (to protect the environment)
12. Society not reacting to dishonest actions
13. Non-existence of political will for effective management of shared resources
14. Failure by both the government and the administration to adopt their proper role in resource management
15. Deficient legislation
16. Lack of respect for EC Directives
17. Lack of legislative enforcement
18. Insufficient sanctions for environmental damage
19. Lack of co-ordination and co-operation
20. Lack of consideration of ecological consequences of decisions
21. Insufficient means for natural resources management
22. Insufficient funding for natural resources management
23. Inadequate training and staffing levels
24. Strong tendency towards the privatisation of public land (short-term private gain over-ruling long-term public interest)
25. "Easy" money-making, resulting in environmental damage
26. "Needs satisfaction" driving human action
27. Single-minded management of water resources
28. Ecological knowledge not sufficiently taken into account and not leading to action
29. Substantial lack of public ecological awareness
30. Public failing to adopt its proper role in resource management
31. Country not alert to environmental deterioration
32. Lack of environmental education
33. Insufficient voluntary participation in nature conservation
34. Lack of ecological education in schools
35. Underdeveloped NGOs
36. Fragmentation in NGOs
37. Lack of broad knowledge on environmental/wetland values
38. Lack of basic ecological knowledge at all levels
39. Lack of specific ecological knowledge

(Pergantis, 1992)

TABLE 3: PHYSICAL THREATS TO GREEK WETLANDS

1. Exploitation of natural forests and clearing of gallery forests
2. Uncontrolled burning of reedbeds or encroachment on reedbeds
3. Pollution from industrial, agricultural and domestic sources, including some from international sources, and pesticide and fertiliser runoff.
4. Herbicide application to floating vegetation
5. Drainage and reclamation
6. Decreased water levels which expose the land to overgrazing, which degrades the riparian vegetation or prevents natural regeneration, and can also result in trampling of nests and chicks.
7. Access roads which increase hunting pressure
8. Wilful destruction or robbing of nests and disturbance during breeding or nesting periods
9. Eutrophication
10. Siltation caused by erosion from surrounding areas
11. Tourism, especially water sports
12. Construction of roads and buildings
13. Intensive aquaculture which has resulted in the introduction of exotic species
14. Destruction and removal of sand dunes through sand and gravel extraction.
15. Use of sandy areas for motorcycle and car racing.

(Valaoras, 1992)

effectively neutralised through action to involve and educate the public or constituency, if only in order to motivate other stakeholders into taking the appropriate decisions. Economic factors account for 5 of the causes, while half of them reflect institutional failures; some of the latter categories, together with pollution causes, are amplified in Table 4.

Although drainage for agriculture has been the major direct (tertiary) cause of wetland loss, threats to wetlands are often indirect, caused by proposals to divert the water supplying the wetland for alternative use, or political purposes. Examples of the former include Botswana's Okavango Delta and the Sudd wetlands on the White Nile, but developed countries can also provide examples such as the Everglades in Florida, which is suffering from both physical and chemical stress (Postel, 1993).

Examples of deliberate political action include the almost irreversible destruction of the mangrove swamps of the Mekong Delta by the defoliant Agent Orange, and the draining of the wetland home of the 'Marsh Arabs' in Iraq, but far more subtle are the political manoevrings around the water resources of the Middle East, India and Bangladesh, the USA and Mexico; problems have become so widespread that the word 'Hydropolitics' has been coined.

It would be fruitless to rehearse further examples of wetland loss throughout the world, since recent publications have already provided chapter and verse in this respect. However, it is appropriate to note just one or two summaries from a recent comprehensive work, 'Managing Mediterranean Wetlands and Their Birds'. This is the proceedings of the symposium held in Grado, Italy (Finlayson et al, 1991), in which Hollis summarises concern over:

> 'a massive loss of wetlands in the Mediterranean. Wetland loss and degradation continue and even the internationally important sites in the region are threatened. This loss and degradation are rooted in social, economic and political processes. These operate behind a chimera created by the immediate causes of wetland loss whilst the apparent causes, such as the oft-quoted agricultural intensification and tourism etc. are merely the outward expression of the underlying factors.'

Likely Effects of Wetland Loss

The proceedings of the Grado symposium provide further information to answer many of the questions posed by the OECD project launched in 1988 on 'Overcoming Impediments to the Integration of Environmental Considerations into Economic Development':

1. What is the present state of the wetland resource stocks, and what trends are apparent?

TABLE 4: TYPE AND SOURCE OF FAILURE

	TYPE OF FAILURE	SOURCE OF FAILURE.
1.	**Pollution Externalities**	
1.a	Air pollution, off-site	Excess levels of sulphur and nitrogen, causing loss of species diversity.
1.b	Water pollution, off-site	Excess nitrogen and phosphorous from sewerage and agricultural sources; some industrial (toxic) pollution
1.c	Water pollution, on-site	Agricultural and recreational pressures
2.	**Public Goods-type Problems**	
2.a	Groundwater depletion	Over-exploitation; on- and off-site use of surface water supply; diminution of wetlands water supplies.
2.b	Congestion costs, on-site	Recreation pressure on wetland carrying capacities
3.	**Intersectional Policy Inconsistencies**	
3.a	Competing sector output prices	Agricultural price-fixing and subsequent land requirements
3.b	Competing sector input prices	Tax breaks or outmoded tax categories on agricultural land; tax breaks for housing or industrial usage; tax breaks on forestry capital; low interest loans to farmers; conversion subsidies (drainage, fill, flood protection, flood insurance); subsidies on other agricultural inputs; and research and development biased towards intensive farming methods.
3.c	Land use policy	Zoning; regional development policies; direct policies favouring conversion of wetlands; agricultural set-aside schemes; waste disposal policies.
4.	**Counterproductive Wetlands Policies**	
4.a	Inefficient policy	Policies that lack a long-term structure.
4.b	Institutional failure	Lack of monitoring and survey capacity, poor information dissemination, non-integrative agency structures.

(Jones, 1991)

2. If the state of the stocks is deteriorating, why should society be concerned? is the loss of these wetlands necessarily a bad thing, or should we be prepared to accept some wetland losses if the potential gains from alternative land uses are greater?
3. If the loss of these wetland services is, on balance, a negative result, what is the source of the problem?
4. What options exist for improving the situation?

(Turner and Jones, 1991; OECD, 1992)

Jones analysed the trends of wetland loss (60,000ha every year in the UK during the 1970s and 1980s, owing largely to conversions to agricultural use). He then briefly considered wetland values, concluding that some net social losses were occurring in the OECD because of the way in which wetlands were being managed; in some cases, economic development would actually be enhanced by arresting wetland degradation. A distinction between wetlands that were 'critical natural capital' and those that could be termed 'substitutional natural assets' could be made. Various types of market and intervention failures were identified and appear here in Table 4 (Jones, 1991).

Natural Recovery or Re-creation
Many of the likely effects of wetland loss are implied in Table 1 in a generic form; specific examples are listed in Table 5. The overall impression is that treating wetlands as a 'free good' is not only ignoring their existing value to the economy, but is rapidly undermining one of the basic elements of the global equilibrium on which Humankind depends. One, furthermore, that has taken centuries or millennia to develop naturally; it has yet to be shown that artificial re-creation of these systems can fully replace their intrinsic value. A strong presumption for conservation before re-creation should therefore be a primary principle in decision-making (Gardiner, 1994a).

The precautionary principle of sustainable development, applied to the slow recovery and practical unrecreatability of many wetlands, coupled with the unknown quantum of their contribution to biodiversity and the local or global ecological carrying capacity, dictates that they be treated as critical natural capital. While many of the tradeable values identified may not be applicable in the UK, the intrinsic value must be taken into account, perhaps as part of a source/sink balance for the UK's carbon cycle to satisfy the need for intra-generational equity implied by the Brundtland definition of sustainable development as that which meets the needs of the present without compromising the ability of future generations to meet their own needs (Brundtland, 1987).

The concept of 'no net loss' (of wetlands) practised in North America has found some problems in application, including one that is of general

TABLE 5: SPECIFIC EXAMPLES OF THE VALUE OF WETLANDS
from the text of *Waterlogged Wealth* (Maltby, 1986)

Genetic Conservation. In the 1960s, scientists from the International River Research Institute in the Philippines screened almost 10,000 varieties looking for a gene giving resistance against grassy stunt virus, a disease then devastating rice crops. Eventually, just two seeds of *Oriza nirana*, and Indian wild rice, were found to have the resistant genes. Researchers returned to the wild habitat to collect more but found none with the correct gene. Today, all the world's rice crop contains genes from those two wild rice seeds.

Water treatment. In Waldo, Florida, United States, sewage purification by flooded cypress groves is calculated to be 60% cheaper than comparative mechanical and chemical methods.

Nutrient removal. In many developed countries, agricultural pollution and sewage causes nitrate levels in rivers, lakes and groundwater to reach levels dangerous to health and to cause eutrophication of marine and freshwaters. Plants such as duckweed can remove up to 67% of such nitrogen in their growth, and water hyacinth up to 92%.

Freshwater Fisheries. The Inner Delta of the Niger River in Mali produces about 100,000 tons (fresh weight) of fish per year and supports many of the villages in the area.

Productivity. The marshy Niantic River in Connecticut, United States, yields an annual scallop harvests greater than that of prime beef on an equivalent area of grazing land.

Firewood. The mangrove Sundarbans of India and Bangladesh provide a commercial crop of firewood as well as 80% of the fish caught in the Ganges-Brahmaputra estuary.

Coastal fisheries. The value of shrimp fisheries in Thailand has been put at $2,000 per hectare a year.

Nurseries for ocean fisheries. Two-thirds of the fish caught worldwide are hatched or spend part of their life-cycle in tidal areas.

Tourism. Some 5 million Americans spend over $638 million a year visiting wildfowl refuges in the United States. 250,000 nature lovers visit Texel Island in the Netherlands annually, and wildlife safaris to the Okavango Swamp in Botswana are worth $13 million a year.

Flooding prevention. On the Charles River in the United States, the value of wetlands preventing serious flooding has been put at £13,500 per hectare a year.

Coastal protection. At Brisbane International Airport, Australia, the cost of planting 51,000 mangrove seedlings in 1981 was $228,271, that is $4.50 per plant.

Energy and carbon dioxide storage. Peatlands cover 500m2 per hectare and are important 'sinks' of carbon, locking it up in dead plant matter. It has been estimated that organic soils store 500 times the carbon released from burning fossil fuel each year. Their destruction could accentuate the greenhouse effect.

Wildlife habitat. Chinese wetlands are home to over 90% of the endangered Siberian cranes. In the United States, 35% of all rare and endangered animals are wetland species.

Of all the commercially important North Sea populations, 60% of the brown shrimp, 80% of plaice, 50% of sole and nearly all of the herring are dependent on the shallow waters of The Netherlands' Wadden Sea at some part of their life history.

significance concerning substitution. Pastor and Johnson (1992) studied the impact of cumulative loss of wetlands on water quality in central Minnesota, USA) using statistical data related by a geographical information system. It was found that both spatial extent and spatial distribution of wetlands affected water quality, and each affects different parameters. The significant finding for regulatory agencies and land managers was that the spatial distribution of wetlands is more strongly correlated with many of the major water quality parameters; 'no net loss' regulation or negotiation may not achieve no net decline in the ability of the system to absorb pollution. In other words, the carrying capacity of the catchment may not be maintained solely through recreation of the same area of wetland lost to development.

For the decision-maker, sustainability depends on knowing what comprises the environment, how it functions and what are its various limits of exploitation: its 'carrying capacity' for human uses. What is its natural regeneration or recovery potential? How 'recreatable' are wetlands, if at all? Can their functionality be achieved otherwise? Wetlands have been diminished down the centuries; the question is whether death is coming by the thousand cuts already delivered, or by the cut to be delivered today. Since nature takes life at its own pace, the answer may only be apparent to our children's children; they are unlikely to know when the fateful cut was made, only that it was their forebears who made death inevitable through lack of effective action.

Sustainable development is about avoiding this unhappy situation; more than that, it is about so altering the trajectory of our resource management that future generations will be able to enjoy an improved quality of life on Earth. Whatever the apparent reasons for wetland loss, it is suggested that this trajectory will spring from a new calculus for decision-making which is based on new perspectives of ecology and social science (Newson, 1992) and according to Feldman (1991) a new theory of justice embracing natural resources as well as people.

Such a theory will need to reflect increasing public concern over the fact that issues between the natural and built environment are increasingly decided by reference to monetary value. Without this framework, should society expect, for example, civil engineers trained as arbitrators to base their decisions over culverting of a watercourse on the likely ecological potential of the river corridor or the possible implications of planning permission for a change in use of adjacent land? And yet this is already happening; is it cut No. 999, or ? What is the relationship between incremental loss of wetland to local development and accumulative loss of 'biogeophysical' carrying capacity for humankind, but calculated within local rather than global boundaries to satisfy the intra-generational equity principle of sustainable development.

Approaches to Sustainable Management

Globally, the challenges appear to be largely political and institutional. The effects of over-population and poverty in developing countries, together with institutional weaknesses that have made poor use of intervention from developed nations and multi-national companies, have recently been recognised by the World Bank. In its 1993 Policy Paper on Water Resources Management, three problems in particular are identified:

- fragmented public investment programming and sector management, that have failed to take account of the interdependencies among agencies, jurisdictions and sectors
- excessive reliance on overextended government agencies that have neglected the need for economic pricing, financial accountability and user participation and have not provided services effectively to the poor
- public investments and regulations that have neglected water quality, health and environmental concerns

To improve the capacity of countries to manage water resources more effectively, 'a balanced set of policies and institutional reforms should be sought that will harness the efficiency of market forces and strengthen the capacity of governments to carry out their roles.' (IRBD, 1993).

A test of this capacity of governments to carry out their roles is whether 'wise use' is made of their country's wetlands. A Conference at Caghari (Italy) in 1980 identified 'wise use' as involving establishment of comprehensive wetland policies based on a nationwide inventory of wetlands and their resources. The next Conference at Groningen (The Netherlands) in 1984 adopted a recommendation which identified Action Points for Priority Attention in the field of wetland conservation. Its annex listed national measures which would promote wise use.

Acknowledging the work of the International Union for the Conservation of Nature and Natural Resources' Wetlands Programme Advisory Committee, the 1987 Regina (Canada) Conference of the Parties to the Ramsar Convention adopted the following definition:

'The wise use of wetlands is their sustainable utilisation for the benefit of mankind in a way compatible with the maintenance of the natural properties of the ecosystem'.

'The Conference noted that sustainable utilisation is 'human use of a wetland so that it may yield the greatest continuous benefit to present generations while maintaining its potential to meet the needs and aspirations of future generations'. The natural properties of the ecosystem were defined as 'those physical, biological or chemical components such as soil, water, plants, animals and nutrients, and the interactions between them'.

'This definition draws attention to the maintenance of not only internationally important wetlands, but all those which bring benefit to humankind, even if only on a local scale. The reference to 'natural properties of the ecosystem' provides a clear insight into what natural functions need to be maintained by wise use. The components that support a wetland often originate well outside its boundary. Similarly, the benefits that wetlands provide to society are frequently manifest beyond the wetland. Wise use of wetlands therefore often requires that appropriate conservation measures be taken beyond the boundary of the wetland. For example, ensuring a continued supply of water of appropriate quality may require soil conservation measures in the headwaters, minimal upstream diversions of river water and the protection of watercourses from industrial pollution' (Hollis et al, 1988a; 1988b).

In the United States of America, the concept of 'Wise Use' is being applied to floodplain management, which is defined as a decision-making process seeking wise use of floodplain lands and waters (Thomas, 1994). According to Thomas, wise use occurs when floodplain activities are compatible with both the flood risks to human life and property and the risks to floodplain natural functions posed by human activities.

Before a proposed activity is located in a floodplain, decision-makers (who are unfortunately not always aware of either flood risk or floodplain natural functions) must determine whether it is possible to reduce the risk of flooding to a politically and socially acceptable level. If not, the proposals are unlikely to constitute wise use, but even if risks can be reduced to an acceptable level, it must then be determined whether the proposed activity will allow for maintenance of the natural functions of the floodplain. If not, then the proposed activity is unlikely to be a wise use, even if the risk to human resources is acceptable. *Wise use (of floodplains) is any use or set of uses compatible with the risk to human and natural resources.* (Thomas, 1994)

To help the decision-maker, Thomas identifies a range of strategies and tools which can be used to determine the 'best mix' (Table 6), which is the combination of human and natural resource strategies and tools most suitable for the uses of a given floodplain. All proposed floodplain activities must be considered in the context of their potential impacts upon river basins and coastal reaches as whole natural systems, the alternative and compatible uses, and the hierarchy of government units (Thomas, 1994).

TABLE 6: STRATEGIES AND TOOLS FOR FLOODPLAIN MANAGEMENT

Strategy A. Modify Susceptibility to Flood Damage and Disruption

1. Floodplain Regulations
   a) State regulations for flood hazard areas
   b) Local regulations for flood hazard areas
      1) Zoning
      2) Subdivision regulations
      3) Building codes
      4) Housing codes
      5) Sanitary and well codes
      6) Other regulatory tools

2. Development and Redevelopment Policies
   a) Design and location of services and utilities
   b) Land rights, acquisition and open space use
   c) Redevelopment
   d) Permanent evacuation

3. Disaster Preparedness
4. Disaster Assistance
5. Floodproofing
6. Flood Forecasting and Warning Systems and Emergency Plans

Strategy B. Modify Flooding

1. Dams and Reservoirs
2. Dikes, Levees and Floodwalls
3. Channel Alterations
4. High Flow Diversions
5. Land Treatment Measures
6. On-site Detention Measures

Strategy C. Modify the Impact of Flooding on Individuals and the Community

1. Information and Education
2. Flood Insurance
3. Tax Adjustments
4. Flood Emergency Measures
5. Postflood Recovery

Strategy D. Restore and Preserve the Natural and Cultural Resources of Floodplains

1. Floodplain, Wetland, Coastal Barrier Resources Regulations (see Strategy A.1.b(1-6) above)
2. Development and Redevelopment Policies (see Strategy A.2(a-d) above)
3. Information and Education
4. Tax Adjustments
5. Administrative Measures

England, Wales and the National Rivers Authority

In England and Wales, the area of the United Kingdom over which the National Rivers Authority (NRA) has jurisdiction, it is probably fair to say that echoes of the concerns expressed by the World Bank can be seen as driving some of the current institutional changes. A major challenge is to provide a model of sustainability involving appropriate administration of land use and economic planning, legislation and multi-functional decision-making. The organisations and remits involved in administration of wetland areas include:

Ministry of Agriculture, Fisheries & Food/Welsh Office	Policy
Departmentment of Authority	Legislation Committments
National Rivers Authority	Operational Authority with Conservation Duties
Internal Drainage Boards	Operational Authority
English Nature/Countryside Commissions	Site Designations, Site Management
Voluntary Bodies (RSPB Wildlife Trusts)	Site Management
Wildfowl and Wetlands Trust	Conservation (wildfowl) advice

In the 80s, models of 'Integrated River Basin Management' and multi-functional projects were successfully exported by the then 'water authorities'. There are already signs that a significant element of a sustainability model built on the co-ordination of land use with management of the water environment involving 'catchment management planning', is similarly providing a useful model abroad. Similar models have been developed in New Zealand and Australia, based in the former on Regional Councils which combine land use planning with water and soil conservation and management. Such models are in general accord with Agenda 21 from the Earth Summit at Rio (Gardiner, 1994a). However, there are several powerful influences which bear on decision-making within the model:

(i) Functions and 'Value for Money'

To maximise its effectiveness as Guardian of the water environment, the NRA needs to be in a position to influence decisions which affect the hydrological cycle. However, the resources available are limited, so the question is how to use them to greatest effect? The traditional, functional approach to water management often resulted in curiously insular and defensive attitudes over some curious and (what are now seen as) indefensible practices. Judging from the literature, this phenomenon is or was common to most countries.

While 'single-function attitudes prevail, the benefits of any proposed activity must more than match the costs for that particular function - and the measurement of wetland benefits for any function is not an easy task. However, there are many cases in which a robust multi-functional case can be made for action, where a single function might fail to find sufficient justification. A range of issues which tends to suffer from this syndrome was identified; Table 7 shows some of them with a few examples which show morphological sensitivity in the responses, and the distinction in terms of wetland protection and creation.

The NRA's process of change from such functionalism to a more multi-functional approach has been underway for longer in some functions than others, and until recently relied rather more on individuals prepared to champion the cause than on institutional instruction.

(ii) Catchment Management Plans

The advent of catchment management planning (CMP) in the NRA provided the institutional instruction. The CMP programme facilitates both multi-functional working internally and an improved focus for external influence on development planning and control externally (Slater et al, 1994). A CMP can be seen as a form of strategic environmental assessment (SEA), isentifying and supporting the framework for environmental assessment (EA) of projects. It has been suggested (Gardiner, 1994a) that such influence on the activities of others can offer a route to sustainable development for the water environment, and hence value for money in the long-term.

This is particularly true when the influence is preventing problems rather than curing them. However, the financial benefits of a 'cure' (persuading a developer to rehabilitate a site, for example) may be more apparent than the value of working with development planners to prevent the degradation happening in the first place. Just as the seduction of major capital works ('Solving the Problem') tends to attract funds away from the area of development planning and control - and is easier to justify funding by virtue of the readily calculable benefits involved - so, in turn, the obvious benefits of development control (a mixture of prevention and cure) may attract funding away from influence on development planning (Figure 1), which is focused on prevention of future problems.

Despite marked progress towards rectifying the resourcing imbalance

TABLE 7: FUNCTIONAL AND MULTI-FUNCTIONAL RESPONSES TO CHANGE

ACTIVITY WITHIN CATCHMENT	IMPACT ON THE WATER ENVIRONMENT	DISCIPLINE AFFECTED	TRADITIONAL ENGINEERING RESPONSE	MULTIFUNCTIONAL RESPONSE
Urbanisation	Lowered base flows	WR/BIO/FISH/CONS	Enlarge channel	Source Control of runoff using swales, permeable pavements etc. **Storage ponds** can control quantity and quality of runoff if **reedbed technology** is employed. The quality of water infiltrating cannot be ignored as it may affect groundwater quality of a known aquifer. Establishment of a **multistage channel** can accommodate increased discharges while maintaining a low flow width. This will create a more aesthetically pleasing environment eg. a **wetland** can be created on the higher berm, or it can be used for recreational purposes.
	Increased flood flows	FD	Line channel with concrete or similar armour layer. Sheet piling or concrete on both banks.	
	Decreased infiltration and lowered groundwater levels through an increase in hardstanding	WR/LAND		
	Increased sediment load initially	WQ/BIO/FISH/CONS		
	Increased discharge leading to erosion (especially in noncohesive glacial sands and gravels)	CONS/FISH/BIO/WQ/GEO		
	Overwide channel encourages deposition of urban silts	CONS/FISH/BIO/FD/GEO		
	Increased surface water discharges (diffuse pollution source)	WQ/CONS/FISH		
Widening/ Deepening	Overwide channel - deposition of silt.	CONS/LAND/FISH/BIO/GEO/WQ	Continually dredge channel to maintain hydraulic capacity. Stabilise banks using sheet piles/concrete.	Allow sedimentation using deflectors to create berm - low flow width later stabilised by vegetation, creating **marginal habitat**. A **two-stage channel** can be created by retaining original width but widening banks, improving aesthetics and recreation/ landscape.
	Deepened channel - unstable banks - rotational or slab failure.			
	Habitat diversity diminished.			
	Lack of depth of flow under normal conditions leading to reduced DO levels.			
Bank Protection	Loss of natural bank profiles. **Loss of marginal habitats.**	CONS/LAND/GEO	Toe boards, Sheet piling, sandbags	The planting of reeds, geotextile stretch fencing, gabions or rip-rap, willow spiling. These will establish a **marginal habitat** that will protect the bank.

EA Co-ordinators, NRA Thames Region (1992).

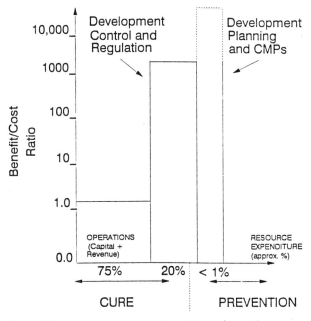

Fig. 1. Possible value for money for NRA related to expenditure

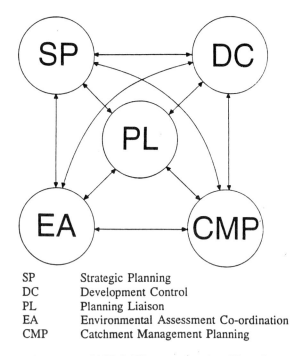

SP   Strategic Planning
DC   Development Control
PL   Planning Liaison
EA   Environmental Assessment Co-ordination
CMP   Catchment Management Planning

Fig. 2. Major elements of NRA Thames Region Planning

inherited by the NRA, it could be argued that the thrust of expenditure remains at odds with the actual value for money to be gained. This undoubtedly reflects the fact that value for money is an insufficient arbiter between different activities, and - again from personal experience - there can be no doubt that the strength of the NRA as a regulator is drawn strongly from its knowledge and experience as an operator.

(iii) The Catchment Planning Team

Nevertheless, it is clear that the protection of existing wetlands from development pressures will depend far more on the influence of the NRA on land use, development planning and control than on its operational activities. That is not to belittle the very considerable achievements of the NRA in creating and rehabilitating wetlands in its operational role. Once an add-on activity to a main function such as flood defence, wildlife habitat rehabilitation and creation has grown in the NRA to become a highly valued activity in its own right.

This welcome change has done much for the credibility of the NRA. It undoubtedly acts as a role model for others (for example, the joint NRA/Thames Water Utilities Ltd. wetland creation at Pinkhill Meadows, Farmoor) and informs and strengthens the NRA's negotiating role in development planning and control.

The value of this dual role of operator and regulator, while not always easy to manage, is seen as a key principle to be carried into an environmental agency; it will do much to ensure real protection of the water environment in general and wetlands in particular. The current 'market testing' process provides an opportunity to review output performance measures and resources in order to ensure that appropriate resources are committed to obtaining value for money through activities aimed at prevention as well as cure.

In practical terms, this objective would be supported by ensuring that the catchment planning and planning liaison activities are closely associated within each region. To complete the model, these teams could operate within a planning team framework which includes dedicated resources for influencing local authority development plans, co-ordinating the authority's negotiations with developers and others over individual planning proposals, which requires expertise in environmental assessment.

Such a team (Figure 2) will be ideally placed to promote consistency in the aims, objectives and standards of guardianship of the physical environment externally as well as internally through catchment plans. By preventing problems at source, through a process of environmental appraisal of development plans, strategic and project environmental assessment, and direct negotiation over the issue of land drainage consent, it can complement the work of pollution control, while co-ordinating with them and indeed all other functions to maximise benefits for the water environment through effective influence on development planning and control (Newson, 1991).

Measurement of the effectiveness and productivity of such a team can be achieved with a report sheet as shown in Figure 3. In terms of value for money, if not potential output performance measures which could reflect 'ground truth', the following results show what is achievable from a regional team with these characteristics:

Making a Difference: Influence on 178 major development schemes: NRA Thames Region Technical Planning Team 91/92
- high level of environmental damage averted
- total asset negotiated over £32.5 million
- environmental assets negotiated £11.6 million
- 50km of river enhancements
- 3km of river saved from culverting
- £5 million per person per year productivity

These assessments involve some judgement as to what would have happened without the involvement of the NRA; in the Author's view they are a realistic and possibly rather conservative assessment of the **difference** achieved by that involvement. There is the significant point that the 'high level of damage averted' may in itself be worth considerably more than the assets negotiated, if only because such prevention applies to a high percentage of proposals received, whereas the asset negotiations have only been recorded where assets in excess of £1000 were achieved (hence the figure of only 178 schemes above). Wetland protection, enhancement and creation is a significant element of this activity.

It was felt that pressures on wetlands and other resources associated with the water environment were significantly alleviated merely by taking a strong stance as guardian; consultants began to advise their clients to approach the NRA team as early as possible in their scheme feasibility stage. On several occasions, pre-feasibility discussions for projects of national significance were requested under conditions of strict confidentiality. This condition can be most trying for an 'open' organisation, especially when extended for a long period (over two-and-a-half years, on one occasion), but in the Author's opinion the benefits of such arrangements to both human and natural communities outweigh the costs.

A major aspect of 'value for money' concerns the use of environmental economics, in particular contingent valuation methods (CVM) such as public willingness to pay or accept change. The technique has been extensively used with success abroad, for example in

```
┌─────────────────────────────────────────────────────────────────────┐
│                 Technical Planning : Reading Office                 │
│                Development control/Environmental assessment         │
│             Situation Report for period April 1992 to September 1992│
│                                                                     │
│   Title & Nature of works    _____         │
│                                                                     │
│   Proposed by  _____   Stage of negotiation  _____        │
│                                                                     │
│   File Ref.  _____   Grid Ref. _____  Status [MR/NMR]    │
│                                                                     │
│   River system  _____   Location  _____        │
│                                                                     │
│   Length of river _____  Estimated cost of scheme _____     │
│                                                                     │
│   Planning Engineer _____   Environmental Coordinator _____   │
│                                                                     │
│   Description of works, concerns and impacts:                       │
│                                                                     │
│   NRA Departments consulted: (please circle as appropriate) Projects│
│   Survey, Modelling Geomorphology, Conservation, Fisheries, Landscape,│
│   Recreation, Biology, Pollution, Groundwater, Water resources, River Ops.│
│                                                                     │
│   Hydraulic/Environmental treatment:                                │
│   - Surveys requested / carried out  _____          │
│                                                                     │
│   - Flood mitigation measures  _____          │
│  ┌─────────────────────────────────────────────────────────────┐    │
│  │ - Mitigation/enhancement (please circle the appropriate numbers)│
│  ├─────────────────────────────┬───────────────────────────────┤    │
│  │ 1. protection of existing   │ 7. restoration of natural     │    │
│  │    river corridors          │    sinuosity                  │    │
│  ├─────────────────────────────┼───────────────────────────────┤    │
│  │ 2. river corridor (beyond   │ 8. treatment of bed through   │    │
│  │    maintenance strip)       │    channel narrowing/creation │    │
│  │                             │    of a low-flow width        │    │
│  ├─────────────────────────────┼───────────────────────────────┤    │
│  │ 3. maintenance strip (the   │ 9. treatment of bed through   │    │
│  │    higher score is for both │    substrate reinstatement    │    │
│  │    banks)                   │                               │    │
│  ├─────────────────────────────┼───────────────────────────────┤    │
│  │ 4. two-stage channel        │ 10. treatment of bed by       │    │
│  │    treatment                │     formation of pools and riffles│ │
│  ├─────────────────────────────┼───────────────────────────────┤    │
│  │ 5. general landscaping      │ 11. structures/any other      │    │
│  │    (outside channel)        │     consentable works         │    │
│  ├─────────────────────────────┼───────────────────────────────┤    │
│  │ 6. treatment of banks       │ 12. environmental treatment   │    │
│  │    (morphology, landscape,  │     of flood storage areas    │    │
│  │    ecology).                │                               │    │
│  └─────────────────────────────┴───────────────────────────────┘    │
│   Length of river enhancement:                       _____ kms     │
│                                                                     │
│   Length of river saved from culverting (if appropriate) ____ kms   │
│                                                                     │
│   Environmental damage averted: A. very high; B. High; C. Moderate; D. Local│
│                                                                     │
│   Technical Planning staff time input:   -engineering  _____ days   │
│                                          -environmental _____ days  │
│                                          -other staff  _____ days   │
│   Evaluation of assets promoted by NRA influence:                   │
│   a) engineering assets (model/survey)         £_____            │
│   b) environmental survey                      £_____            │
│   c) value of engineering works                £_____            │
│   d) value of enhancements                     £_____            │
│                                                                     │
│   Savings to Developers                                             │
│                                                                     │
│   Description  _____       £  _____            │
└─────────────────────────────────────────────────────────────────────┘
```

Fig. 3.

New Zealand (Harris, 1994). Despite some notable results which clearly indicate public concern over issues such as low river flows and channelisation, the CVM has yet to find any general acceptance by the UK Treasury as justification for action.

Further Issues for the NRA - some examples
The successful management of wetlands is highly dependent on their geological, hydrological and topographical structure. A current NRA research project is seeking a 'hydrotopographical' classification of British wetlands to guide managers in their decision-making. As an example, a clear distinction can be made between wetlands which are fed from unconfined aquifers and those which rely on river flooding; such a basic difference can be central to the quality of decisions over appropriate management, and underlines the importance of developing a robust classification.

The basic issue of classification is complemented by the issue of 'need'. How can the intrinsic value of wetlands be identified sufficiently to afford protection and encourage conservation, enhancement and rehabilitation, if not creation? Apart from the scale of ecological science required, this immense challenge involves environmental ethics, social and political awareness, effective legislation, inter-organisational co-operation etc. At ground level, it depends on human resources; not just numbers of people and time, but also (crucially) empowerment of individuals - a neat phrase summarising a welter of educational, institutional, pschological and other factors which determine how well legislation is enforced, and whether negotiation is successful or takes place at all.

There are a number of areas in which the NRA faces serious issues in carrying out policy. Several raise the question of the NRA's remit, not only in protecting wet grasslands for example but also in promoting source control (management of rainfall-runoff at or near the point of rainfall). Both of these fall into the multi-functional trap.

Once involved, however, what line is the NRA to take over, for example, implementation of Environmentally Sensitive Areas? The Levels and Moors ESA is undoubtedly of great potential value to the water environment, but as part of its 'Water Level Management Strategy' should the NRA support expenditure probably in excess of £¼ million (shared with English Nature and the Ministry of Agriculture, Fisheries and Food) on engineering activities to safeguard farmers who do not wish to be included, or offer to pay them full compensation for complete rehabilitation of whole hydrological units (the moors) by the 'natural' approach of turning-off the pumps?

The issue is over implementation of the sixth item of the NRA's policy for The Somerset Levels and Moors (NRA, 1992). This states that the NRA will adopt a presumption in favour of positive water level management for nature conservation on SSSIs, and in other appropriate areas where there is general agreement. Joint guidelines from the Ministry of Agriculture, Fisheries and Food, NRA and English Nature on 'Water Level Management Plans', to be published in June 1994, will address this difficult issue.

Again, what is the NRA's response to a developer who claims that his development of a holiday village for 3000 people can be successfully achieved on drained moor which has been agriculturally improved but with drainage channels of SSSI status between 75 and 300 metres apart? Clearly a case for an Environmental Assessment, which will need to focus on the operational state, when a mix of horseflies, midges, children and rubbish spell doom for the development or the SSSI; buffer zones of adequate dimensions would constrain the development unacceptably. In such a case, the strength of the partnership between NRA and local authority (over their shared vision expressed in the development plan) would be tested.

Partnerships are increasingly being recognised as the key to progress on the ground. There are many examples in addition to those already mentioned, including the action plan to safeguard the watermeadows and wildbrooks of Sussex (Whitbread and Curson, 1992) and the Pevensey Levels study (Hart and Douglas, 1993). It was the rejection in 1978 of a proposal by Southern Water Authority to drain the Amberley Wild Brooks in Sussex which was the first public inquiry held under the Land Drainage Act 1976 and an important test case for wetland protection (Hall, 1978; Penning-Rowsell 1978).

Examination of just a few examples underlines the need for a common language, understanding and vision between central government, local authorities and the environmental agencies, if not all the stakeholders involved. This common language can be sponsored by the drive towards sustainable development, if agreement is sought over ecological and social principles and sustainability criteria (Gardiner, 1994b; Clarke and Gardiner, 1994). The recent Thames Region initiative over the practical definition of environmental capacity, as applied to Cotswold Water Park and proposed further mineral extraction, has been used as a test vehicle for this approach (Gardiner 1994c).

Involvement of stakeholders can be rewarded by the sort of public initiative sponsored by the 'Urban Streams Restoration Project', begun by Anne Riley of the Californian Water Resources Department in Sacramento, or its equivalent run by Anne Jensen in Australia's Murray-Darling basin. The success of both programmes has had much to do with the empowerment of the individuals involved. While this new paradigm is unfolding, the pressure on wetlands continues; importing world-wide examples of best practice is needed given the urgency of the situation.

The post-event discussions on the Mississippi floods provide a further forum for such best practice to be identified, acknowledging that the 'best mix' for the Mississippi floodplain may well differ from the Ganges, Brahmaputra and Meghna floodplains, which may show some modest differences from the Thames. The need for communication through all sectors of society and across continents is urgent, since there is little time if this vital element of our environment is to be protected.

Conclusions

If the world's wetlands are under such pressure and have already been reduced by 50% in area, much the same can be said about rainforests. Taken together, they show incremental but rapid and massive destruction and deterioration of ecological systems that have taken eons to evolve. Apart from the many known benefits provided by wetland systems (see Tables 1 and 5), there are probably many more unknown benefits to the humankind, in addition to supporting the 'local' populations and playing a vital role of unknown magnitude in maintaining global homeostasis. Viewed from space, it may seem that Mankind is engaged in a dangerous game of pushing to the limit the carrying capacity of the Earth's biological resources which maintain homeostasis.

Portrayal of the global scale of wetland losses with magnificent photography in 'Wetlands in Danger' (Dugan, 1993) underlines the significance of the issues. The concluding passage, 'Challenge to Conservation' uses headings such as population growth; uninformed development; economic growth; using water; rising sea levels; changing communities; priorities and quantification; demonstrating incentives; communication of information; wetland programmes; the role of the NGOs; institutional capacity; the Ramsar Convention; development assistance policies, and the integrated approach. These headings summarise the scope of current understanding, underlining the issues brought out in this paper.

A summary of the conclusions to this paper would include:

Causes of wetland loss:
- Increase in population/need for self-sufficiency or overseas earnings
- Poor perception of wetland values compared with alternative uses of land
- Weak institutional protection and enforcement
- Fragmentation of technical and scientific interests
- Dominance of agriculture, industry and infrastructure interests

Effects of wetland loss:
- Reduction of biodiversity and genetic variety within species
- Reduction of carbon store/sink

- Disruption of ecological pathways
- Loss of buffer zone functions (flood attenuation, water quality etc)

Recovery/Re-creation:
- Evolution over millennia - effectively unrecreatable
- Recreation of full functionality yet to be proved
- Policies of 'no net loss' to include systemic effects of location

Approaches to sustainable management of wetlands:
- Agreed classification and inventory (shared between agencies)
- 'Capacity Building' ensuring effective institutional influence
- Adoption of sustainable development philosophy and practice in decision-making
- Achieve stakeholder consensus over principles and criteria for sustainability
- Environmental/Social Appraisal of policies, plans and strategies
- Multi-functional integration/co-ordination through SEA and project EA
- Team dedicated to prevention and mitigation of damage to the physical environment through co-ordination of influence on land use and development planning based on catchments - complementing pollution control based on legislation
- Identification of strategies/tools to facilitate 'wise use' and 'best mix' eg. floodplains
- Use of environmental economics and economic incentives
- Public involvement, education and ownership of the issues

As David Bellamy comments in the Introduction to 'Wetlands in Danger':

'Peat-forming wetlands, coral reefs and some marine plankton are today the only things that perform the vital function of keeping the gases of the atmosphere in balance. They are the true 'lungs' of the Earth. The world's peatlands alone cover more than 2.3 million square kilometres with 330 billion dry tons of organic matter. If burned or drained and so opened up to slow oxidation, they would release 500 billion tonnes of carbon dioxide , almost double the amount present in the atmospheric greenhouse today..... It has been calculated that 1 hectare of tidal wetland can do the job of US$123,000 worth of state of the art waste-water treatment, and many communities are now recreating wetlands to cleanse their waste.....there are lights at the end of tunnels, but time is running out for so many crucially important sites and for the world at large, for without wetlands the biosphere cannot continue its vital work.'

For England and Wales, much has been done to protect remnant wetlands from further erosion and loss. However, there are serious issues of value, recreatibility, criticality and sustainable management of the resource at both local and national scales, to be addressed. Every day, in the planning offices of the NRA, proposals are received seeking to replace wetlands with other land uses. It is so easy to lose such environmental assets, and yet it has been shown that they can be protected and enhanced with relatively little extra effort.

Acknowledgement
The views expressed in this paper are those of the Author, and not necessarily shared by the NRA. The Author's thanks go to Les Jones, Regional General Manager of NRA Thames Region, for his kind permission to publish this paper.

References
Brundtland, G.H. (1987). Our common future. Oxford University Press, Oxford.

Brennan, A. (1994), Environmental Literacy. Environmental Values 3(1), The White Horse Press, Cambridge.

Clarke, M.J. and Gardiner, J.L. (1994). Strategies for handling uncertainty in integrated river basin planning. Paper to the Int. Conf. on Integrated River Basin Development, Wallingford, U.K.

Dugan, P. (1993). Wetlands in danger. Mitchell Beazley, London, 187pp.

Feldman, D.L. (1991). Water resource management: in search of an environmental ethic. Johns Hopkins University Press, Baltimore, MD.

Finlayson, C.M., Hollis, G.E. and Davis, T.J. (Eds). Managing Mediterranean wetlands and their birds. Proc. Symp., Grado, Italy. IWRB Spec. Publ. No.20, Slimbridge, UK.

Fordham, M., Tunstall, S. and Penning-Rowsell, E.C. (1989). Choice and preference in the Thames floodplain: the beginnings of a participatory approach? In Marchand, M. and Udo de Haes, H.A., (eds) The people's role in wetland management: proceedings of the international conference at Leiden, The Netherlands, pp.791-797.

Gardiner, J.L. (1994a). Sustainable development for river catchments. J.IWEM, (In press).

Gardiner, J.L. (1994b). Developing flood defence as a sustainable hazard alleviation measure. In Gardiner, J.L., Yevjevich, V., and Starosolszky, O. (eds) Defence from floods and floodplain management; procs. of NATO Advanced Study Institute, Budapest, Hungary.

Gardiner, J.L. (1994c). Capacity Planning and the Water Environment. RSPB 2nd. National Planners' Conference, RSPB, Sandy, Beds.

Hall, C. (1978). Amberley Wild Brooks. Vole 7, pp.14-15.

Harris, B. (1994). Personal communication.

Hart, C. and Douglas, S. Pevensey Levels study. National Rivers Authority, Southern Region.

Holdgate, M. (1990). Changes in perception. In Angell, D.J.R., Comer, J.D. and Wilkinson, M.L.N. (Eds). Sustaining earth: response to the environmental threat. Macmillan, London.

Hollis, G.E., Larson, J.S., Maltby, E., Stewart,R.E. and Dugan, P.J. (1988a). The wise use of wetlands. Procs. of 3rd Meeting of the Conference of the contracting parties to the convention on wetlands of international importance, especially as waterfowl habitat; Regina, 1987. Ramsar Convention Bureau, Bland, Switzerland pp.263-268.

Hollis, G.E., Holland, M., Maltby, E. and Larson, J. (1988b). The wise use of wetlands. Nature and Resources, 24(1), pp.2-13.

Hollis, G.E. (1991). The causes of wetland loss and degradation in the Mediterranean. In Finlayson, C.M., Hollis, G.E. and Davis, T.J. (Eds). Managing Mediterranean wetlands and their birds. Proc. Symp., Grado, Italy. IWRB Spec. Publ. No.20, Slimbridge, UK. pp.83-91.

IRBD, (1993). Water resources management. The World Bank, Washington, 140pp.

Jones, T. (1991). In Finlayson, C.M., Hollis, G.E. and Davis, T.J. (Eds). Managing Mediterranean wetlands and their birds. Proc. Symp., Grado, Italy. IWRB Spec. Publ. No.20, Slimbridge, UK. pp.220-226.

Lovelock, J. (1989). The ages of Gaia. Oxford University Press, Oxford.

Marchand, M. and Udo de Haes, H.A., (eds) The people's role in wetland management: proceedings of the international conference at Leiden, The Netherlands, 872pp.

Meadows, D.H., Meadows, D.L. and Randers, J. (1992) Beyond the limits. Earthscan, London. p64.

Newson, M.D. (1991). Catchment control and planning: emerging patterns of definition, policy and legislation in UK water management. Land Use Policy, 8(1), Butterworths, pp.9-16.

Newson, M.D. (1992). Water and sustainable development: the 'turn around decade'? Journal of Environmental Planning and Management, 35(2), pp 175-183.

OECD (1992). Market and government failures in environmental management: wetlands and forests. OECD, Paris, 82pp.

Pastor, J. and Johnston, C.A. (1992). Using simulation models and geographic information systems to integrate ecosystem and landscape ecology. In Naiman, R.J. (ed) Watershed management: balancing sustainability with environmental change. Springer-Verlag, New York, pp.324-346.

Penning-Rowsell, E.C. (1978). Proposed drainage scheme for Amberley Wild Brooks, Sussex: benefit assessment. Middlesex University, Flood

Hazard Research Centre, London.

Pergantis, P. (1991). Strategic schemes to improve wetland management in Greece. In Finlayson, C.M., Hollis, G.E. and Davis, T.J. (Eds). Managing Mediterranean wetlands and their birds. Proc. Symp., Grado, Italy. IWRB Spec. Publ. No.20, Slimbridge, UK. pp.226-231.

Postel, S. (1993). Last oasis: facing water scarcity. W.W. Norton & Company, New York, pp.60-67.

Purseglove, J. (1988). Taming the flood. Oxford University Press, Oxford. p55.

Ramsar Bureau (1989). The Ramsar Convention. Ramsar Bureau, Gland, Switzerland and Slimbridge, U.K.

RSPB (1994). Wet grasslands - what future? Royal Society for Protection of Birds, Sandy, Beds.

Scottish Natural Heritage, (1993). Sustainable Development and the Natural Heritage: the SNH approach. Scottish Natural Heritage, Redgorton, Perth.

Slater, S., Marvin, S. and Newson, M.D. (1994). Land use planning and the water environment: a review of development plans and catchment management plans. Working paper No.24, Dept of Town and Country Planning, University of Newcastle upon Tyne.

Turner, K. and Jones, T. (1991). (Eds) Wetlands: market and intervention failures. Earthscan Publications, London, U.K. 202pp.

Valaoras, G. (1991). Greek wetlands: present status and proposed solutions. In Finlayson, C.M., Hollis, G.E. and Davis, T.J. (Eds). Managing Mediterranean wetlands and their birds. Proc. Symp., Grado, Italy. IWRB Spec. Publ. No.20, Slimbridge, UK. pp.262-267.

World Bank Policy Paper WR Management

Whitbread, A and Curson S., (1992). Wildlife drying up. Sussex Wildlife Trust,

Estuaries and wetlands: function and form

J. PETHICK, Director, Institute of Estuarine and Coastal Studies, University of Hull

The management of tidal wetland areas demands a careful evaluation of their location and importance within the wider estuarine system. This paper reviews some of the work carried out by the author on the estuaries and associated inter-tidal wetlands of the east coast of the UK. Wetlands, defined here as including inter-tidal mudflats and salt marshes, are described as an integral part of estuarine development and hydrodynamics and, more specifically, represent a morphological response to tides, waves and sea level changes. The location and extent of inner-tidal deposits is attributed to the overall estuarine tidal processes while the extent of salt marshes and in particular the salt marsh/mudflat boundary is a response to wave energy. The importance of the functional interaction between estuary and inter-tidal deposits and between salt marsh and mudflat is stressed as are the interruptions to these interactions provided by such factors as reclamation and dredging of estuaries. Wetland management, including restoration and maintenance techniques, can only be effective if these larger scale interactions are understood.

Introduction

The estuaries of eastern England provide a unique series of examples of adjustments to variations in tidal range, wave energy inputs, sea level and sediment supply. These adjustments, both over time and in space, allow an assessment to be made of the relationship between estuarine form and function, a relationship which determines the overall shape of the estuary as well as that of the smaller scale units, such as the mudflats and salt marshes which make up the banks of an estuary. The form, extent and location of these small scale inter-tidal units can only be understood if the larger form function relationships of the estuary are determined. This paper reviews a more extensive study of the estuaries and associated inter-tidal areas of the east coast of Britain. The 13 estuaries in the sample range from the southernmost Medway in Kent, to that of the Dornoch in north east Scotland, encompassing a total coastline of over 1000km (Table 1). The paper outlines the general uniformity of the form and function in this widely spaced sample, but emphasises that the deviations from the general model caused mainly by human interference can be of fundamental importance to the smaller scale relationships between inter-tidal deposits and the wider estuary system.

Two major controls of the location and extent of inter-tidal wetlands are considered. First, it is proposed that a functional relationship exists between depth and length which determines the location of the inter-tidal deposits in an estuary. Second, the well known relationship between estuarine inlet area and tidal prism is extended to show that the areal extent of inter-tidal deposits throughout an estuary is related to tidal prism increment.

Estuarine function and wetland location

Estuaries are essentially different landforms from rivers. It is obvious enough that they experience two-way tidal flows rather than the uni-directional flow of terrestrial streams, yet a more important, but subtle, contrast is the fact that these tidal flows are themselves dependent on the morphology of the channel. In a river, discharges are independent of the form and are determined by the catchment hydrology and any change in channel morphology which introduce a perturbation into the system is soon

Table 1: Tidal properties of selected east coast estuaries

Estuary	Tidal range(m)	Tidal length (km)	Tidal area (ha)
Medway	5.5	55	5794
Thames	5.0	84	18302
Crouch/Roach	4.5	40	1606
Blackwater	4.5	21	5180
Colne	4.2	13	680
Stour	3.6	14	2530
Orwell	3.3	20	1790
Deben	3.2	19	1790
Alde	3.6	27	1020
Blyth	3.0	9	300
Humber	6.0	65	33360
Dornoch	3.4	43	5250

damped out in the downstream direction and not transmitted through the whole channel. In an estuary, however, the tidal prism are uniquely determined by the size of the channel so that any change in the channel is reflected in changes in discharge throughout the system.

Unlike rivers, tidal estuaries tend to maximise energy dissipation so that tidal and wave energy is damped within the channel and not exported. This means that the plan form and the cross sectional shapes adopted by estuaries tend to be quite distinct from those of rivers. In particular the exponential decrease in width inland is a characteristic of most estuaries, while their width-depth ratios tend to be at least an order of magnitude higher than rivers experiencing similar discharges. These morphological adjustments to the tidal inputs tend to maximise frictional drag so that tidal range is reduced to zero at the head of the estuary. Most of this frictional drag is exerted within the inter-tidal zone of the estuary – the mudflats and salt marshes – which are merely the channel banks of the estuary and which consequently adjust to the imposed energy inputs from the tide.

The rate of width decrease in a tidal estuary is the principal determinant of the extent of intertidal sedimentation and therefore one of the most important controls on wetland formation. The rate of width decrease inland can be expressed as :

$$W_x = W_o e^{-x/\sqrt{gD}}$$

Where:

W_x = width at distance x

W_o = width at estuary mouth

x = distance from mouth

D = Mean depth

Since the exponent here includes an expression for tidal wave celerity ($\sqrt{gD}$) this suggests that the plan form of an estuary is related to its tidal inputs. The tidal wave length celerity itself determines tidal wave length (λ) since:

$$\lambda = T\sqrt{gD}$$

and if the estuary length (L) is plotted against the tidal length (λ), that is the distance between the mouth and the upstream tidal limit, for the east coast estuaries (fig 1) it is found that the relationship is linear with a slope of 0.27 and a correlation coefficient of r = 0.93.

This important relationship between tidal wave length and estuary length is a purely morphological one since, as shown above, tidal wave length is determined by estuarine mean depth. Thus the uniformity expressed in fig 1 demonstrates that the length/depth ratio for a tidal estuary tends towards a constant.

The functional significance of this observed form ratio may be that these east coast estuaries, and by implication all estuaries, act as sediment sinks so that their inter-tidal areas extend and their mean depths decrease over time. As the mean depth decreases so the tidal wave length is reduced until it approaches the limit of

$$L \rightarrow 0.25\lambda$$

At this stage the tidal wave becomes fully resonant (Dyer, 1986; McDowell & O'Connor, 1977) which, in theory, will result in a large tidal amplitude at the estuary head, the location of the anti-node, while tidal range at the estuary mouth, the nodal location, is relatively small. This is not observed in any of the east coast estuaries in the sample, which, on the contrary, all exhibit attenuation of the tidal amplitude inland due to decreasing channel width, suggesting that frictional damping has cancelled out the effects of resonant amplification. Since flow rates at the estuary head - that is the anti-node - will be at a minimum and those at the mouth - the node - will be maximum, inter-tidal deposition rates will be increased further landwards into the estuary. This leads to the exponential plan form of the estuary described above and, at the same time, gives a landward increase in the frictional drag of the channel so resulting in a long term steady state between morphology and tidal energy.

The significance of the relationship between tidal wave length and estuary length for wetland management lies in the critical importance of channel depth. Tidal wave length is controlled by the mean depth of the channel and this relationship is extremely sensitive to any imposed natural or artificial changes in channel morphology. It appears that in almost every case such artificial changes lead to depth increases. For example, although reclamation may not alter the channel maximum depth, the removal of inter-tidal areas by reclamation effectively increases the mean depth. Conversely, dredging within a navigation channel may increase the channel maximum depths but, in a wide estuary with extensive inter-tidal zones, this may have little effective on mean depth although in a narrow channel the same amount of dredging may cause dramatic increases in mean depth.

In each case, increases in the mean depth lead to an increase in the tidal wave length and cause the estuary to deviate from the stable relationship outlined above. For the east coast estuary sample these deviations are small : fig 1 indicates that only 4 of the 13 cases lie outside the 95% confidence limits to the best fit regression line, but they are extremely significant for the inter-tidal areas of the estuary and, more particularly, for their management. In most cases it appears that reclamation of inter-tidal wetlands is responsible for the deviations to the best fit relationship shown in fig 1. The crucial role of inter-tidal wetlands in maintaining the channel mean depth means that any interference with their location or extent must lead to instability, a factor examined in more detail later in this paper. It is interesting, however, to note here that the most pronounced and the only positive anomaly is that of the Dornoch estuary in north east Scotland – where very little inter-tidal reclamation has taken place. Then explanation for this anomalous morphology may lie in the fact that the Dornoch is the only estuary in the sample to have experienced a steady fall in sea level over the past 3000 years, a phenomenon which has continued to the present day (Smith, Firth, Turbayne, & Brooks, 1992). This means that the mean depth of the estuary has decreased over time, leading to a tidal wave length which is much shorter than would be predicted for the estuary length. The extensive inter-tidal deposition in the inner Dornoch and the spits and tidal deltas forming at its seaward extremity appear to be morphological adjustments to this instability.

It is also important to note that the estuary of the R. Crouch has been considered here together with its tributary, the R. Roach. The combined tidal lengths (Table 1) of these two estuaries is plotted in fig 1. Plotted as single estuaries they show as extremely large anomalies to the general trend - a result which suggests that bifurcating estuaries should be treated as a single unit.

Estuarine function and wetland extent

The role of the inter-tidal deposits of an estuary in defining its overall plan shape, through the length/depth ratio described above, can be further defined by the well known relationship between tidal prism and the cross sectional area of the estuarine inlet. Since tidal prism in an estuary is itself largely defined by the extent of the inter-tidal area this relationship is fundamental to wetland development. Moreover, the extension of this relationship to include increments of the tidal prism landward of all cross sections within the estuary, indicates that plan-form and cross section form are only part of a three-dimensional whole.

Many authors have commented on the relationship between the cross sectional area of the mouth of an estuary or tidal basin and the tidal prism (Bruun, 1978; Bruun & Gerritson, 1960; Escoffier, 1940; Gao & Collins, 1994; O'Brien, 1931; O'Brien & Dean, 1972). One criterion on which this relationship could rest centres around the attainment of a critical current velocity for a stable inlet which, according to most authors, is around 1.0ms-1, although Gao & Collins (1994) suggest some deviation from this. The relationship is often expressed, following O'Brien, 1931; O'Brien & Dean (1972) as a linear regression between $\log_{10}A$ and $\log_{10}P$ (where A_m is the cross sectional area(m^2) and P the tidal prism (m^3)):

$$A_m = CP^n$$

Values of C and n for a world-wide sample of tidal inlets given by Gao & Collins, (1994), suggest that C can vary over a wide range from 0.0034 to 0.235 and n from 0.84 to 1.05.

The sample of east coast estuaries shown in Table 1 follows this relationship closely, a significant linear relationship between A_m and P ($r = 0.90$) is demonstrated (fig 2). The magnitude of the exponent $n = 1.20$ in this relationship lies just outside the range for the sample given by (Gao & Collins, 1994) but the intercept $C = 5.1 \times 10^{-7}$ is far smaller and probably reflects the restricted tidal areas of these east coast estuaries.

The excellent fit of the data to the regression line in fig 2, indicates the general uniformity of morphology throughout the east coast estuaries but here, as in the relationship between length and depth (fig2), these anomalies are critical to any wetland management programmes for specific estuaries. The deviations from the best fit regression to the P and A data set are indicated in fig 2 where the estuaries of the Medway Blackwater and Blyth are shown as lying on or beneath the 95% confidence limits and that of the Crouch and Dornoch as on or above this limit. (Note that the Crouch is again considered here in conjunction with its tributary, the R. Roach. It is the combined tidal prism of both estuaries which is plotted in fig 2.) Most of these anomalies appear to be due to artificial changes induced in the estuary, mainly through reclamation of inter-tidal wetlands and this is discussed in detail below. However, it is interesting to note that the Dornoch again appears as an anomaly not because of artificial interference in its system but probably due to sea level fall which has resulted in a tidal prism which is smaller than would be predicted from its inlet dimensions. The extensive deposition at the mouth of this estuary, mentioned above, forming the Dornoch spit and the Morrach More, appears to be an adjustment to this instability.

Although most authors have emphasised the relationship between tidal prism and cross sectional area, some have used tidal discharge(e.g. McDowell & O'Connor, 1977) as a more appropriate process variable. In the case of an estuary, tidal discharge and cross sectional area are not merely related at the mouth, but throughout the whole estuary system, since a critical discharge, and therefore flow velocity, exists for each cross section in the channel. This relationship between tidal discharge and channel morphology was evaluated by Langbein (1963) in his study of the hydraulic geometry of a small estuary. Langbein proposed that the exponents in the hydraulic geometry equations for an estuary were quite distinct from those for rivers, in particular the sensitivity of width and the relatively conservative response of depth to changes in discharge. These equations were:

$$w = aQ^{0.71}$$
$$d = cQ^{0.24}$$
$$v = kQ^{0.05}$$

Where:

w = width

d = depth

v = velocity

Q = discharge

a, c, k are constants

These relationships demonstrate that, as the tidal prism decreases landwards in an estuary, there is a rapid exponential decrease in width but a much more subdued change in depth and velocity. Applying these equations to the Blackwater estuary, the present author showed (Pethick, 1994) that changes in tidal prism and therefore in discharge, brought about by sea level rise could explain the temporal changes in estuary width over the past 150 years which have resulted in major loss of inter-tidal wetland. Over the past 20 years alone salt marsh loss in this estuary has amounted to 200ha or 23% of the total area (Burd, 1989). Moreover the application of the Langbein equations to predicted changes in sea level rise in the future could represent a management tool with which to plan for changes in channel morphology, including such concepts as managed retreat in order to maintain both channel efficiency and the area of ecologically important wetland in the estuary.

WAVES AND INTER-TIDAL PROFILES

The tidal dynamics of an estuary are described above as resulting in the extent and location of inter-tidal deposits and in their cross sectional form. The profile of the inter-tidal zone, however, and, crucially, the location of the salt marsh/mudflat boundary is probably as much a product of wind wave processes as it is of tidal action.

Research in the Blackwater estuary (Pethick, 1992) has identified the nature of the relationships between wave action and inter-tidal profile form. Using a series of wave recorders coupled with long term profile elevation monitoring, the effect of extreme wave events on mudflat morphology has been shown to be similar to that of beaches, that is a flattening during storm events and a subsequent recovery to give a steeper profile during low energy periods. This work has identified two quite distinct processes acting in the inner and outer estuary. In the outer estuary, ocean waves are able to propagate to approximately 10km inland; these ocean waves have an open ended distribution, that is larger waves are associated with longer return intervals, so that the inter-tidal zone is subjected to high magnitude events. During such high energy wind wave events, mudflats profiles tend to erode at their upper margins and the sediment transferred to the low tide margin. This process during extreme wave events can even result in the horizontal retreat of the salt marsh edge as the mudflat erodes into it. This results in a flatter and wider mudflat profile over which wave attenuation is increased and thus represents an efficient morphological adjustment to storm events. The mudflat and salt marsh of these outer areas of the estuary are thus seen to exist as a single geomorphic and sedimentological unit, analogous to that of the beach/sand dune system. The profile of this combined unit responds to wave events in a dynamic manner with the saltmarsh–mudflat boundary in particular reflecting periodic variations in energy inputs. Observations during one extreme storm event in the Blackwater, estimated to have a 30 year return period, showed that the salt marsh suffered both horizontal and vertical erosion but this was balanced by accretion on the lower mudflat. Recovery of the salt marsh towards its initial pre-storm morphology was subsequently observed to take place. Extrapolating the measured rates suggest that a full recovery from this 30 year event will take 10 years, indicating that storm effects merely interrupt but do not reverse, the long term development of marshes in these exposed estuarine areas.

In the inner estuary a quite distinct set of process elements and morphological responses was observed. In the first place ocean waves do not affect these areas so that all waves are generated within the estuary itself and are fetch limited - that is the maximum wave height is determined, uniquely, by the width of the water surface over which a given wind is blowing. This produces a distinctive wave height distribution, with a marked upper limit determined by the maximum fetch length and means that the mudflat-salt marsh system in these inner areas experiences a constant wave energy input to which it responds. Thus the mudflat profile develops a slope and width which is capable of fully attenuating the maximum wave height over a fixed distance, at which point marsh development commences. The result is that in these inner areas the mudflat-marsh boundary is constant, as opposed to the fluctuating boundary of the outer estuary, and consequently the more rapid accretion on the marsh surface results in a discontinuity which takes the form of a small marsh cliff. Despite their appearance these marsh cliffs are not erosional but depositional forms and reflect the balance between wave energy and frictional drag provided by the inter-tidal profile. The marshes in the Colne estuary, for example, are bounded by a marked 0.5m cliff but historic map and chart evidence together with direct observation over the past 15 years has shown that the position of these micro-cliffs have remained constant for at least 150 years (Burd 1989).

Long term development

The relationships between components of estuary morphology such as length, depth and width, and its tidal prism and discharge alter continuously as deposition occurs within the system. On the east coast of Britain, approximate present sea level was attained some 6000 years ago; since then all the east coast estuaries have experienced rapid deposition resulting in profound modification of their overall morphology and dynamics - mostly involving inter-tidal wetland development The discussion given above allows the general form of this development throughout this 6000 year period of the Holocene to be proposed. As deposition proceeded in the typical east coast estuary, so mean depth decreased and a tendency towards a resonant standing tidal wave was initiated. The balance between the tidal forces and deposition processes at this stage maintains the estuary morphology in steady state, mainly due to the controlling presence of the inter-tidal areas which define both the overall estuary shape and its cross section profile.

This morphological steady state represents a long term average condition of the estuary; over shorter time intervals maintenance of a stable morphology depends on the continuous adjustment of inter-tidal morphology via the processes of deposition and erosion. In the east coast estuaries, sediment is derived principally from marine sources and is carried into an estuary by tidal and residual currents. The estuary acts as a sediment sink in the early stages of its development but, as a steady state with its tidal forces is approached, sedimentation rates are reduced and may reverse to give short term erosion.

This change from sediment sink to sediment source, is a response to changes in channel cross section. Immediately after the initial rapid Holocene rise in sea level the drowned valleys were relatively deep and wide. Since tidal wave celerity is given by $\sqrt{gD}$, and since mean depth in these wide, deep channel cross sections would have been significantly greater at high tide than at low tide, the tidal wave progression would be more rapid at high water than at low and thus the wave form would have tended to become asymmetric, giving a flood tide dominant velocity. This in turn would result in a net sediment input to the estuary so that it acted as a sediment sink. As sedimentation continued, so the elevation of the inter-tidal zone would have increased relative to the maximum depth of the subtidal channel so that the mean depth of the channel decreased, leading to the progression to tidal resonance noted above.

The continued inter-tidal deposition in the estuary would have caused the cross section to change to that of a deep central channel bounded by high inter-tidal banks in which the mean depth at high water would be less than that at low water. In such a channel the crest of the tidal wave tends to drag relative to the trough and an ebb-tide dominance is set up (Dronkers, 1986). Consequently a negative sediment balance would have developed in the estuary which would have become a net sediment source rather than a sink. Loss of sediment from the inter-tidal zone increases the mean depth, however, and tends to return the channel cross section to its former flood tide dominant shape. Thus over a long time period, perhaps measured in hundreds of years, the estuary oscillates around a steady state condition with alternate erosional and accretional phases characterising its inter-tidal zone.

These phases in development may be seen in the east coast estuaries where flood dominant estuaries such as the Humber contrast with those exhibiting an ebb dominance such as the Blackwater or Crouch. The explanation for the difference in the regimes may be that both are responding to oscillations around the steady state as described above, but alternatively it may that the relatively high demand and low supply in the northern estuaries has retarded their development while the smaller demand and higher supply in the southern estuaries such as the Medway, Thames and Blackwater has allowed them to achieve a more mature morphological stage. In this case it is more likely that these more mature southern estuaries may exhibit alternating erosional and depositional phases. In the case of the Medway it has been proposed (IECS, 1993)that the quarrying of inter-tidal mudflats in the last century has merely accelerated one of these erosional phases and that signs of incipient recovery are now apparent in the estuary as a flood dominant tidal system develops.

Sea level changes and wetland development

Sea level changes in these estuaries also plays an important part in their long term morphological development. An increase in sea level causes an initial increase in channel mean depth and therefore in flood asymmetry, so that sedimentation rates increase, depth decreases and stability is maintained. This positive feedback mechanism is identical to that which maintains the critical depth/length ratio described above and is a normal part of the Holocene estuarine response. The Blackwater estuary appears to typify such a response, here erosion of the inter-tidal and deposition of the eroded sediments in the sub-tidal channel is leading to a change from ebb dominant flow to flood dominant flow. This may promote an increase in imports of sediment from marine sources and lead to a gradual recovery of the inter-tidal zone so that natural wetland restoration can take place as the inter-tidal migrates landwards.

The implications of this long term development for wetland management are critical. Erosion of the inter-tidal zone may be seen as a necessary stage in the maintenance of the steady state morphology over the Holocene period and should not be arrested using, for example, artificial defences. Flood defences constructed or strengthened due to sea level rise, for example, merely confine the channel and prevent the landward migration of inter-tidal wetland - the phenomenon known as coastal squeeze. Similarly, dredging the channel may alter the natural processes although prediction of the outcome of dredging is complex . In the Thames estuary for example, dredging in the outer estuary has resulted in a deep narrow central sub-tidal channel but has not altered the high inter-tidal banks. This has resulted in an ebb-tide dominance in which net loss of sediment means that the navigation channel is self maintaining. In the Gravesend reach, however, the same volume of dredging in a smaller cross section has removed inter-tidal as well as sub-tidal sediment so that the channel has become relatively narrow and deep and a flood dominance set up in which sedimentation is rapid and requires constant maintenance dredging to keep it open.

The intertidal morphological unit

Inter-tidal wetlands have been shown in this paper to owe their location and extent to estuarine tidal dynamics while their profile, and the location of the mudflat-marsh boundary upon it, is determined by wind wave processes. The importance of viewing the mudflat-marsh as a single unit is stressed in this analysis. Salt marshes cannot exist in isolation but require the low energy conditions provided by a wave attenuating mudflat surface. Equally, however, outer estuary mudflats can only respond to high magnitude wave events by eroding landwards into the salt marsh surface allowing them to widen and providing a short term source of sediment. Where a backing salt marsh is removed, for example by reclamation, mudflats are denied both this landward extension and the short term sediment source and consequently suffer more extreme erosion during storms. They are also slower to recover from such storms, in some cases failing to make a complete recovery before the next high magnitude event occurs so that a slow deterioration in the inter-tidal zone takes place, marked by a fall in mudflat elevation especially at the upper margins. This can have serious repercussions for flood defence embankments as well as inter-tidal biology, but even more important is the overall change in the estuary dynamics which can occur. A general decrease in mudflat elevations throughout an estuary can lead to increased tidal prism and tidal discharges resulting in more extensive erosion - a positive feedback which appears to have occurred, for example in the Medway Estuary. Here the removal of salt marsh sediment for use in the brick making industry led to increased tidal current velocities as well as increasing the available water surface are so that internal wave generation increased. The result was that the initial anthropomorphic disturbance was amplified with the loss of 1512 ha of marsh from an initial area of 2157ha and the lowering of mudflat surfaces by up to 2.0m (Kirby 1990, IECS, 1993).

In the inner estuary the interaction between the two components of the inter-tidal unit takes a different form. Here salt marsh development again requires that a fronting mudflat is present in order that internally generated waves are attenuated but, since there is an upper limit to wave height, the mudflat itself does not experience the periodic erosion and recovery phases involving the marsh sediments as in the outer estuary. This means that the inner estuary mudflats are not directly dependent on the salt marsh, but nevertheless there is an indirect relationship here. The presence of salt marshes in the inner estuary limits both the tidal prism and the wave fetch so that the tidal and wave energy inputs to the mudflat are kept to a minimum. Where salt marshes are missing or are restricted, mudflats either suffer progressive erosion as in the case of the Medway cited above, or experience low or zero accretion rates.

The estuaries of the Alde and Blyth in Suffolk both demonstrate this latter effect. In the Blyth, for example, reclamation had, by 1840, reduced the former inter-tidal extent by 1100ha and formed a narrow parallel sided channel. Collapse of the inner reclamation embankments in the 1940's resulted in the return of 250ha to the inter-tidal (Beardall, Dryden, & Holzer, 1991). Within this area, high tidal discharges and velocities coupled with wind waves generated over the large expanse of open water at high tide inhibit accretion rates so that no salt marsh development has taken place. In the Alde estuary a similar situation has developed but here it is progressive reclamation of the outer estuary together with the onshore movement of the Orford Ness spit that have reduced the outer channel while leaving a wide inner inter-tidal area in which, again, inhibition of mudflat accretion has kept salt marsh development to a minimum.

In both these cases it appears that the lack of salt marshes has both contributed to, and is a consequence of, the low mudflat accretion rates, consequently additional factors must be sought in order to explain the initiation of this circuitous process. It was suggested above that the location and extent of inter-tidal areas within an estuary was a response to gradual changes in the tidal dynamics. The temporal development envisaged is one in which both process and form adjusted together, a relationship first described by Wright, Coleman, & Thom, (1973):

> Neither the channel morphology nor the tidal properties can be explained solely in terms of each other, though the two are mutually dependent. Simultaneous coadjustment of both process and form has yielded an equilibrium situation in which further adjustment is nonadvantageous (Ibid p 28).

Once this mutual response has been reached any major interference with the system, such as that which has taken place in the Blyth and Alde, can only be accommodated by morphological changes across the entire system. In the case of these two estuaries such adjustments are prevented by the flood embankments which effectively constrain any further changes. In the case of the Medway, the initial growth of inter-tidal deposits during the Holocene took place in an estuary with two mouths, one each side of the Isle of Grain. One of these, the Yantlet channel, was blocked by embankments during Roman times after salt marsh growth in the main estuary had reduced the tidal prism to a minimum. The recent removal of salt marshes has meant a huge increase in tidal prism but the mouth cross section is still restricted to a single channel between Grain and Sheerness and it may be that natural regeneration of marsh cannot take place unless the original mouth morphology is restored. It is significant to note that the Medway plots as a negative anomaly to the P-A regression line (fig 2) suggesting that its inlet dimensions are too small for the tidal prism. However, if the inlet dimensions are increased to include the whole of the former Yantlet Channel mouth then the estuary falls directly on the best-fit regression line for the east coast sample.

This effect of reclamation on estuarine morphology and in particular on the management of its inter-tidal wetlands, is considered in more detail in the following section.

Wetland reclamation and retreat

Any changes to the estuarine morphology, whether natural or artificially induced, are transmitted throughout the system as a whole and particularly affect the inter-tidal zone. Many of these changes and their results have already been documented above. For the east coast estuaries, reclamation has constituted the most significant interruption to the system in the past, although other factors such as navigation dredging, training walls, restriction of sediment supply from defended cliffs and port and harbour construction are locally important. In the future it may be that the managed or even unmanaged retreat of reclamation defences may constitute the major interference in the estuarine system. In this section some of the effects of reclamation are reviewed in the context of the east coast estuaries.

	Y-direction	X-direction	Examples
Positive anomalies	Estuary too long	Estuary too shallow	Dornoch
Negative anomalies	Estuary too short	Estuary too deep	Blyth, Crouch, Blackwater

Table 2a: Deviations (>95%) from the best fit regression to the estuary length–tidal wave length relationship.

	Y-direction	X-direction	Examples
Positive anomalies	Inlet too wide	Tidal prism too small	Crouch, Dornoch
Negative anomalies	Inlet too narrow	Tidal prism too large	Blyth, Medway, Blackwater

Table 2b: Deviations (>95%) from the best fit regression to the tidal prism – inlet dimension relationship

Reclamation of the inter-tidal zone in the east coast estuaries as a whole has removed large areas from the active estuarine channel. Although figures for the whole of the east coast are not available, the total loss of salt marsh for 7 of the 13 estuaries in the sample shown in Table 1 amounts to 18680ha, representing 32% of their present estuarine area. Most of this change has taken place over the past 300 years, although some reclamation on this coast was undertaken as long ago as Roman times. The enormous change in the tidal area of east coast estuaries has not only resulted in a significant loss of wetland habitat but also has major repercussions for the dynamics of the remaining inter-tidal areas. These effects can be seen by noting the deviations from the best fit regression lines in figs 1 and 2. In each case a deviation from the regression may be attributed to one or both of two morphological changes. These are summarised in Tables 2a and 2b where examples of positive and negative anomalies which exceed the 95% confidence limits are also shown.

In each case the Dornoch plots as a positive anomaly, a result which has been attributed above to the falling sea level in this estuary rather than any artificial changes. The other anomalous estuaries, however, have suffered extensive reclamation of their inter-tidal areas and it is suggested that this can account for their departure from the stable morphological relationships.

Reclamation can have several effects on the overall morphology of an estuary. These include:

- Reduction of the tidal prism due to widespread reclamation.

- Reduction of the inlet dimension due to seaward reclamation

- Reduction of tidal length due to landward reclamation

- Increased tidal wave length following an increased depth as high inter-tidal areas are reclaimed.

These effects are themselves inter-related, as where reclamation reduces both the tidal prism and increases the mean depth of a channel. This is best shown in the R. Crouch where the extensive reclamation of the Dengie marshes, Foulness Island and Wallasea Island have resulted in a narrow, almost parallel sided, estuary in which the mean depth is too great and the tidal prism to small to allow for morphodynamic stability. In this case, as in many others, the presence of flood embankments prevents any natural adjustments of the estuary towards stability. However, it is interesting to note that where flood embankments have failed in the R. Crouch, as at Fambridge or Bridgemarsh Island, the resulting inter-tidal areas have eroded rather than accreted, resulting in an increase in tidal prism.

The Medway estuary on the other hand has experienced a decrease in its inlet dimensions due to early reclamation, coupled with an increase in tidal prism due to mud-digging as noted above. It seems likely that this estuary may recover from these changes as inter-tidal deposition reduces the tidal prism, although the reduced inlet dimensions may mean that some form of artificial recharge may be needed to initiate this process. The importance of the reduced mouth dimension here may be that it prevents effective ebb tide flows so that tidal water is held back in the inner estuary allowing internally generated wave action to develop and reduce accretion rates.

The estuaries of the Blackwater and Blyth appear to have similar problems to that of the Medway. Constriction of the mouths of these estuaries by reclamation has resulted in an imbalance with their tidal prism. Since the inlet dimensions are constrained by flood embankments, flow velocities have increased in order to provide the necessary tidal discharge into the large inner estuary. In order to realise these velocities on the ebb tide, a hydraulic head has to develop between the inner and outer estuaries so that a longer residence time develops for water in the inner estuary. The resultant wave action reduces inter-tidal accretion rates so that morphological recovery is not possible. As a result, the extensive mudflats in these inner estuaries do not appear to be developing new salt marsh and this has been a major cause of concern among conservationists responsible for the management of these areas. The failure of these estuaries to increase their inter-tidal deposits also means that they remain too deep for the estuary length so that each estuary plots as a negative anomaly on the tidal length – tidal wave length graph (fig 1). In each case these effects are exacerbated by sea level rise which serves to increase the tidal prism without any commensurate increase in inlet dimensions.

Finally, it is interesting to note that where reclamation schemes have cut across large salt marsh creeks in many of these east coast estuaries, the result has been an extension of the beheaded creek along the front of the reclamation embankment, often causing erosion to its toe. This effect is a small scale indication of the balance between tidal prism and inlet dimensions discussed above in its wider estuary context. Removal of the inner portion of a salt marsh creek from tidal action means that a reduction in tidal prism occurs. Compensatory adjustments are possible here however since the remaining salt marsh is not protected by embankments as is normally the case in larger estuary channels and consequently the creek enlarges normal to its original course and the balance is restored.

Conclusions.

The importance of inter-tidal wetlands to the development of estuarine morpho-dynamic stability has been emphasised in this paper. For the wetland manager the converse argument suggests that only if estuarine dynamics are understood can salt marsh and mudflat processes be interpreted correctly and appropriate managerial action taken. Three major conclusions could be drawn from the analysis presented above:

- Salt marshes cannot be created or restored unless adequate mudflats or other forms of wave attenuation are present or can be provided.

- Salt marshes and mudflats define both the estuary tidal wave length and its tidal prism. Any change in the location or extent of such inter-tidal deposits either by reclamation or by managed or unmanaged retreat of existing reclamation areas will result in an imbalance among these estuary properties which may have serious negative impacts throughout the system.

- Attempts to restore or create salt marshes and mudflats should take into account the existing estuary morpho-dynamics. Many estuaries are so narrow and deep due to removal of inter-tidal areas that only extensive retreat would allow them to develop a natural intertidal profile. Other estuaries are so constrained at their inlets as to prevent efficient tidal flows and so inhibit inter-tidal development.

The importance of regarding the estuary as a single functional landform should be central to any management plans. Inter-tidal wetlands define this landform and their location and extent can only be adjusted as part of an integrated strategy.

Acknowledgement: Part of this research was carried out with funding from English Nature and Scottish Natural Heritage; this, and permission to use the data in this paper, is gratefully acknowledged.

References

Beardall, C. H., Dryden, R. C., & Holzer, T. J. (1991). The Suffolk Estuaries. Colchester: Segment Publications for Suffolk Wildlife Trust.

Bruun, P. (1978). Stability of tidal inlets. Amsterdam: Elsevier.

Bruun, P., & Gerritson, F. (1960). Stability of tidal inlets. Amsterdam: N Holland Publ.

Burd, F. (1989). The saltmarsh survey of Great Britain: an inventory of British saltmarshes. No. Research and survey in nature conservation, No.17.). Nature Conservancy Council, Peterborough.

Dronkers, J. (1986). Tidal asymmetry and estuarine morphology. Netherlands J. Sea Research, 20, 117-131.

Dyer, K. R. (1986). Coastal and estuarine sediment dynamics. Wiley. .

Escoffier, E. F. (1940). The stability of tidal inlets. Shore and Beach, 8, 114-115.

Gao, S., & Collins, M. (1994). Tidal inlet equilibrium in realtion to cross sectional area and sediment transport patterns. Estuarine, Coastal and Shelf Science, 38, 157-172.

IECS (1993). Medway Estuary: Coastal processes and conservation. Report to English Nature. Institute of Estuarine and Coastal Studies, University of Hull.

Kirby, R. (1990). The sediment budget of the erosional inter-tidal zone of the Medway Estuary, Kent. Proc. Geol. Assoc., 101(1), 63-77.

Langbein, W. B. (1963). The hydraulic geometry of a shallow estuary. Int. Assoc. Sci. Hydrology, 8, 84-94.

McDowell, D. M., & O'Connor, B. A. (1977). Hydraulic behaviour of estuaries. London: Macmillan.

O'Brien, M. P. (1931). Estuary tidal prism related to entrance areas. Civil Engineering, 1(8), 738-739.

O'Brien, M. P., & Dean, R. G. (1972). Hydraulic and sedimentary stability of tidal inlets. In 13th Coastal Engineering Conference, (pp. 761-780).

Pethick, J. S. (1992). Salt marsh geomorphology. In J. R. L. Allen & K. Pye (Eds.), Salt marshes (pp. 41-62). Cambridge University Press.

Pethick, J. S. (1994). Managing managed retreat. In K. Pye (Ed.), Conf.Procs. Environmental sedimentology, . Reading

Smith, D. E., Firth, C. R., Turbayne, S. C., & Brooks, C. L. (1992). Holocene relative sea-level changes and shoreline displacement in the Dornoch Firth area, Scotland. Proc. Geol. Soc, 103, 237-257.

Wright, L. D., Coleman, J. M., & Thom, B. G. (1973). Processes of channel development in a high-tide range environment: Cambridge Gulf-Ord River Delta, W Australia. J of Geol., 81, 15-41.

Somerset levels and moors water level management and nature conservation strategy

K. W. TATEM, BSc(Hons), CEng, MICE, MIWEM, North Wessex Area Flood Defence Manager, and I. D. STURDY, Somerset Moors and Levels Surveyor, National Rivers Authority, South Western Region

SYNOPSIS. The National Rivers Authority's Somerset Levels and Moors Water Level Management and Nature Conservation Strategy is an initiative by the NRA South Western Region to determine and to provide water levels which will be maintained in selected areas, with the primary aim of benefitting wildlife conservation and sustaining traditional, extensive farming practices on the Somerset Levels and Moors.

1. INTRODUCTION
The Somerset Levels and Moors are within the area of central Somerset, bordered by the Bristol Channel and the Mendip Hills to the north, and the Blackdown, Brendon and Qauntock Hills to the south and west. (Figure 1.)

Much of the land is below the mean sea level of the Bristol Channel and man-made drainage systems are required to drain the area and protect it from flooding.

Within the Somerset moorlands, environmental and topographical features have combined with past agricultural and drainage activities to contribute to the formation of a virtually unique environment which supports a diverse community of flora and fauna.

Until the early 1980's agricultural improvement in the area was encouraged by a number of Land Drainage Acts which were aimed at improving flood protection, reducing waterlogging and enabling agricultural intensification.

The development of water engineering skills combined with improvements to land drainage infrastructure over this period achieved a more controllable water table.

Although the Levels and Moors have retained much of their unique wetland character, the cumulative effects of drainage and changes in land management over the past 50 years are becoming apparent. There has been a dramatic decline in breeding waders over the past 10-15 years. Indeed in certain areas some species of wading birds have already become extinct. Other aspects are giving cause for concern; in particular the average numbers of wintering birds has decreased while the botanical interest of areas such as Tealham and Tadham Moors is undergoing a rapid change. These losses are attributed by the major conservation bodies to the drying out of the moors as a result of the continued maintenance of low water tables.

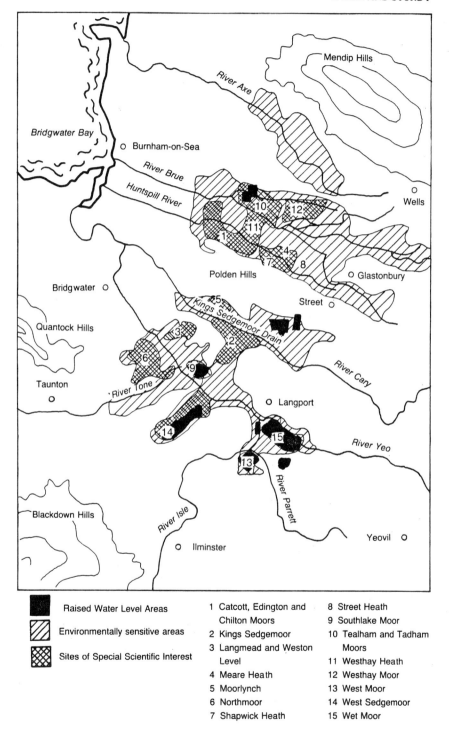

Fig. 1. The Somerset levels and moors

OVERVIEW OF THE SYSTEM

Concern has also been expressed that this drying out process will damage sites of archaeological interest.

Increased awareness of the detrimental effect of low water levels on moorland flora and fauna has led to the designation of Special Status on a number of sites within these areas.

Primarily this Special Status takes 2 forms:

i. Environmentally Sensitive Area (ESA): this is designated by the Ministry of Agriculture Fisheries and Food (MAFF). Within these area, voluntary management agreements are offered to landowners whereby fixed compensation payments are made to owners who manage their land to environmentally friendly prescriptions. Within the ESA 15 selected areas are designated as Sites of Special Scientific Interest (SSSIs). (SSSIs also exist outside the ESA.). Figure 1.

ii. SSSIs:- These are notified by English Nature (EN). Within these areas farming practices are restricted by mandate. Farmers are however compensated for restrictions imposed on their farming activities, on a profits foregone basis. Other designations include scheduled ancient monuments (SAMS).

At an international level the Moors and Levels is a proposed RAMSAR site and a special protection area (SPA) under the EC Directive on Conservation of Wild Birds 1979.

2. SUMMARY OF RELEVANT SCIENTIFIC INFORMATION

A considerable amount of research has been carried out which is of relevance to the decline in the wildlife interest of the Levels and Moors. This work can be conveniently divided into 3 sections: breeding waders, plants and soils. Certain trends are apparent from this work and the key points are outlined over.

Breeding Waders

The habitat requirements of breeding waders are well understood. Each species has preferred areas, snipe, for example, occur in wet, tussocky grassland and require an extended period of wet ground conditions in order to feed. A broad pattern of redshank, curlew and lapwing are seen to occur on progressively drier margins. These species preferences are clearly demonstrated on West Sedgemoor The Royal Society for the Protection of Birds reserve where water levels have been raised to create 400 acres of wetland.

Plants

Recent work, carried out by Silsoe College and the Institute of Terrestrial Ecology on Tadham Moor is particularly relevant.

A multi-disciplinary team has been studying the effects of the height of the water table and the way in which it changes throughout the year in relation to the plants present. Such work has also been carried out in other parts of Europe.

Results so far confirm that the average height of the water table determines the plant species which occur. English Nature has stated that on Tadham Moor the water level is now too low to support the original floristically rich meadows.

Soils
The oxidation of peat soils as a result of lowered water tables and intensive agricultural management is giving growing concern on the Somerset peat moors.

The loss of surface organic matter can lead to the formation of hard unwettable nodules which are extremely difficult to re-hydrate when trying to achieve the normal soft winter conditions.

Studies have indicated that this irreversible drying can cause grass production to fall by 20%.

In addition, the aeolian erosion of this dry powdery surface leads to an alarmingly rapid decrease in land levels and thus the need for even lower water levels. Clearly an unsustainable regime.

It is now widely accepted that reviving and maintaining the former diverse ecology of the moors is inseparabley linked to the provision of higher water tables.

3. THE TWELVE POINTS OF THE STRATEGY
It is the international recognition of the area's wetland value set against the trend of environmental decline that has led the NRA to formulate its Somerset Levels and Moors Water Level Management and Nature Conservation Strategy.

The details of this strategy were determined after the establishment and monitoring of high water level trial areas and after consultation with over 200 other bodies and organisations. Its aim is to work alongside the other initiatives in the area such as the Ministry of Agriculture's Environmentally Sensitive Area Scheme to reverse the decline and to work towards a sustainable balance of all the interests within the Somerset Moors and Levels.

As a result of the consultations and discussions undertaken the National Rivers Authority has adopted the following strategy:

i. The National Rivers Authority recognises the outstanding nature conservation interest of the Somerset Levels and Moors and that this is in decline.

ii. The Authority seeks to restore and maintain the wildlife and landscape of this internationally important wetland, consistent with its given duties, and to conserve the archaeological interest.

iii. The Authority has statutory obligations as regards water management, including the control of water abstraction, discharges, water quality, drainage and water levels.

iv. The Authority will give special consideration to the environmental impact of abstraction and discharges throughout the Levels and Moors.

v. The Authority will review its flood defence practices and take into account the requirements for nature conservation, to ensure sympathetic management within the Environmentally Sensitive Area (ESA). Formal management plans will be agreed with English Nature (EN) over activities which affect Sites of Special Scientific Interest (SSSIs).

OVERVIEW OF THE SYSTEM

English Heritage will be consulted over matters that affect Scheduled Ancient Monuments (SAMs).

vi. The Authority will adopt a presumption in favour of positive water level management for nature conservation on SSSIs, and in other appropriate areas where there is general agreement. Priority will be given to the core areas of SSSIs.

vii. Where raised water levels affect agricultural productivity the Authority will support the introduction of a water level premium on ESA payments and/or Section 15 management agreements with English Nature to offset these costs.

viii. The Authority will liaise with relevant organisations to draw up a list of priority sites where enhanced water levels are required to maintain and restore the nature conservation interest.

ix. The Authority will take action after consultation with the Ministry of Agriculture, Fisheries and Food, English Nature, Internal Drainage Boards and landowners in order to achieve the conservation objectives.

x. The importance of the "withy" growing industry is fully recognised and in implementing its strategy the NRA will seek to accommodate its special requirements.

xi. In implementing the strategy the Authority will take special account of the statutory, practical and financial position of Internal Drainage Boards.

xii. Any changes in strategy must ensure that there is no increase in flood risk to human life, habitation or communications.

The National Rivers Authority has given consideration to those areas of the Somerset Levels and Moors in need of most urgent attention and has adopted the following priority list of areas in which raised water levels will prove most beneficial. These are:

      Catcott, Edington and Chilton Moors
      King's Sedge Moor
      Moorlynch
      North Moor
      Southlake Moor
      Tealham and Tadham Moors
      West Sedge Moor
      Wet Moor   (Figure 1)

A primary aim is to provide core area where shallow winter flooding will create feeding sites for wintering water fowl, and soft ground for wading birds to breed in spring and early summer.

4. **MINISTRY OF AGRICULTURE FISHERIES AND FOOD - (MAFF) AND THE ENVIRONMENTALLY SENSITIVE AREA SCHEME (ESA)**

The Somerset Levels and Moors ESA was established by the Ministry of Agriculture Fisheries and Food (MAFF) in 1987 and covers approximately 27,000 ha.

Its principal aim is to safeguard the traditional, pastoral landscape of the Levels and Moors with its associated wildlife and archaeological features.

The ESA scheme was reviewed in 1991, the main change being the introduction of water level prescriptions to agreement types Tier 1 and Tier 2 and the addition of a Tier 3 and a water level supplement. The purpose of the Tier 3 and the water level supplement is "To further enhance the ecological interests of grassland by the creation of wet winter and spring conditions on the Moors." This is to be achieved through stringent land management restrictions combined with the following water level management prescriptions.

From 1st May to 30 November water levels in the adjacent/peripheral ditches and rhynes must be maintained at not more that 300mm below mean field level and from 1st December to 30 April at not less than mean field level so as to cause conditions of surface splashing. These water level requirements compliment the NRA's strategy particularly well and hence a means by which water levels may be elevated by agreement rather than imposition has been achieved through collaboration between the NRA and MAFF.

5. THE DEVELOPMENT OF A RAISED WATER LEVEL AREA
   Example:- Wet Moor SSSI

Wet Moor SSSI has within its boundaries a fully operational raised water level scheme which produces Tier 3 water level conditions in approximately 650 acres of pastoral grassland.

It is a priority site under the NRA's strategy and is within the Somerset Levels and Moors Environmentally Sensitive Area. The general arrangement of the drainage channels and infrastructure, pre raised water levels is shown on Figure 2.

Fragmented Interest in raised water levels in this area was expressed via the ESA scheme in 1992. Subsequent visits to interested land owners and their neighbours resulted in coherent blocks of land with interest in raised water levels being achieved in approximately 650 acres of the SSSI.

It was at this stage that the NRA were able to consider in detail the options for raising water levels in line with the previously mentioned seasonal freeboards.

To determine the degree of ditch water level elevation required by an area, and indeed by individual field enclosures. There are 3 essential requirements.

   i.   Topographical Information.
   ii. Hydrological/Surface Water Management Information.
   iii. Achieving a hydrological balance.

In Wet Moor both hydrological survey and detailed topographical survey were carried out, enabling the seasonal freeboard of each field enclosure to be determined. Details of the hydrological survey and the topographical survey are given in Figures 2 and 3. This study showed that none of the area being considered for raised water levels met the criteria required. ie Water level elevation was required.

OVERVIEW OF THE SYSTEM

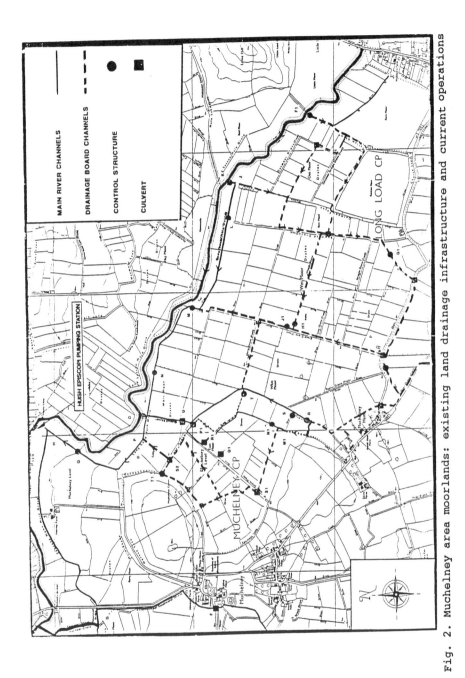

Fig. 2. Muchelney area moorlands: existing land drainage infrastructure and current operations

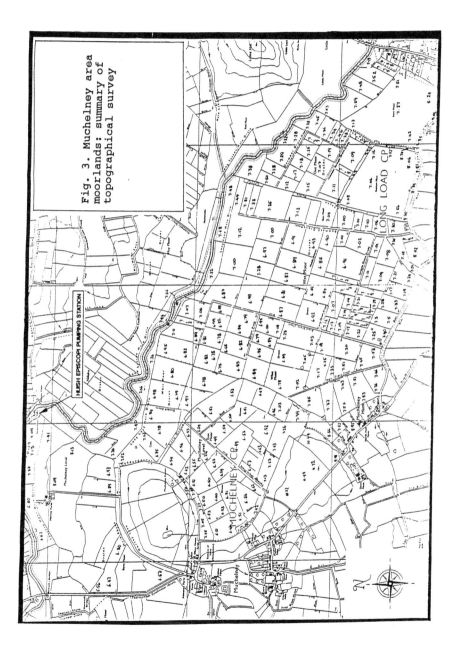

Fig. 3. Muchelney area moorlands: summary of topographical survey

OVERVIEW OF THE SYSTEM

i.   Topography (Figure 3)

Study of the topographical survey revealed that although the area is low in natural relief it is far from level. The difference in land elevation from the east to the west of the site being approximately 1m. (ie 7.5m AOD to the east and 6.5m AOD to the west).

ii.   Surface Water Management (Figure 2)
      (Pre Raised Water Levels)

The moorland of the Wet Moor area is low lying low relief land typical in character of the Somerset Moors and Levels. The landscape is dominated by 2 major arterial watercourses, the Rivers Parrett and Yeo. These are large embanked channels which act as upland water carriers conveying water from the upstream catchments through the low moorlands. In storm conditions the rivers can surcharge their banks and overspill into the adjacent low lying moorland. Here, flood waters are retained until river levels recede below bankfull conditions.

The flood waters may then be evacuated to the River Yeo by pumping at Huish Episcopi Pumping Stations (HEPS). Overtopping of the floodbanks is not an uncommon event. It can occur 2 or 3 times each year. The overtopping usually results in limited flooding of agricultural land and the flood waters are promptly evacuated as river levels and downstream conditions allow.

Within the moorland water is conveyed to and distributed from the main rivers through networks of non-main river channels. These channels serve both as a means of flood evacuation and as irrigation channels for the agricultural area. They are generally small, with extremely shallow bed and water level surface gradients and are often poorly maintained. Flow in these channels is dominated by upstream and downstream water head difference rather than by bed gradient. Many of these watercourses are managed by the local Drainage Board. Those channels which are managed neither by the NRA of the Drainage Board are in private ownership and are managed by the adjacent landowners.

Huish Episcopi Pumping Station (HEPS) is the major control mechanism for surface water management in this area. HEPS evacuates flood water from the area as previously discussed.

It controls water levels for the whole area; maintaining a high summer water level for irrigation and wet fencing purposes and ensuring low water levels during the winter period in order to extend the grazing season and to enable prompt evacuation of excess water which may enter the moor.

The target water levels retained at HEPS are determined by the surface elevation of the lowest land in the moor, at a level of approximately 6.5m AOD. The resulting target water levels held are:

        Summer   6.3m AOD
        Winter   6.0m AOD

When these water levels are compared to the topographical survey it is apparent that large water level to ground level free boards can result in much of the moor during the spring and summer, with the situation being exaggerated even further during the winter months.

This has proved satisfactory in the past for agricultural purposes but without a doubt has damaged the interest of the SSSI.

iii. Hydrological Balance

The achievement of raised water levels is also dependant upon a surplus of water which can be directed to and retained within the desired areas, to counter the natural water losses which occur. The exact volumes of water which pass through the area are unknown. It is however, clear that the majority of excess water eventually passes to Huish Episcopi Pumping Station. Analysis of pump hours at HEPS revealed that substantial quantities of water are evacuated into the river Yeo. The table below summarises the results of this analysis.

Table 1: THREE YEAR AVERAGE VOLUMES OF WATER PUMPED

MONTH	TOTAL HOURS	HRS AT 5.1 cumecs	VOLUME
January	440	147	2,698,920m^3
February	510	170	3,121,200m^3
March	420	140	2,570,400m^3
April	80	26	477,360m^3
May	90	30	550,800m^3
June	100	33	605,880m^3
July	90	30	550,800m^3
August	80	26	477,360m^3
September	80	26	477,360m^3
October	100	33	605,880m^3
November	115	38	697,680m^3
December	225	75	1,377,000m^3

This gives an indication of the magnitude of water which is evacuated from the area. This water is available to achieve and sustain a high water level environment within the required areas.

5.1 ACHIEVING RAISED WATER LEVELS

The presence of a substantial surplus of water which passes through the area, makes the raising of water levels a more viable proposition. After further detailed investigation 2 main options became apparent for providing the area with raised water levels:

i. Raising the water level at the pumping station by reduced pumping.

ii. Strategic retention of water in selected areas. (Terraced Irrigation)

OVERVIEW OF THE SYSTEM

i. **Water Level Control at Pumping Station**

The raising of water levels by maintaining a higher water level at the pumping station appeared the simplest option and least onerous in terms of water level management resources and additional infrastructure. By retaining a revised water level at the pumping station, water levels in the main and adjoining channels would be lifted accordingly and would have provided a "maximum wettable area". There are few additional costs (apart from ensuring ditches are well maintained) associated with this method and initially it appeared to many that this was the logical solution to raising water levels. Upon closer inspection it became clear that problems may occur of this option were employed. These are discussed below.

Selection of Retained Level

The selection of a suitable water level to be retained at the pumping station to service the whole area is a speculative task and must be carried out with a great deal of care. An understanding of the watercourses and topography of the area is essential as the water level ultimately retained must be transmitted to the area where it is required.
The determination of the water level that will be produced in the centre of the moor as a result of the level retained at the pumping station is also an imprecise exercise.
In additon elevation of water levels across the whole area is a non selective excercise resulting in flooding of those who had not agreed to accept raised water levels.

Topography

As previously discussed and demonstrated by the topographical survey, although low in natural relief, the area is by no means flat. A single water level imposed upon the area will therefore produce variable depths of water which may at some locations positively disadvantage both the environment and farming activities. For example, if a summer water level is selected to achieve wet conditions in the largest area possible, large areas of sub-optimum conditions would also be produced from the extremes of both exessive wetness and dryness.

Flood Storage Capacity and Evacuation

Retention of a higher water level at the pumping station will affect all channels which connect to the drainage network of the area. Most importantly the main river channels which are currently managed in the winter to allow prompt evacuation of flood waters will have a delayed effectiveness because of the larger volumes of water which they will retain. In addition, where the "maximum wettable area" is achieved, much of the flood storage capacity for the area will be permanently occupied and an increase in flood risk to roads and settlements in the area may occur.

Agricultural Activity and the Environment

The environmental revival and subsequent maintenance of the Somerset Wetlands is dependant upon the provision of higher water tables produced by supersaturation of the soil during the winter by shallow splash flooding and by generally raised ditch water levels. The practice of traditional farming methods is also essential. Where water levels do not comply with those identified as optimum, the recovery of the wetlands will not be achieved. It is likely with the maintenance of a set level controlled at the pumping station that the worst consequences of both high and low water level management will occur.

Those areas with very low ground levels will be subjected to deep inundation for long periods with farmers unable to achieve access and subsequent degradation of the land. In addition long periods of deep flooding in the spring will produce anaerobic ground conditions and kill off much vegetation.

Those areas with more elevated ground levels will be affected very little by the new water level maintained and current farming activities will continue with the associated slow environmental decline. It is clear therefore that this method of raising water levels should be employed with great caution and it has not implemented at Wet Moor.

ii. Terraced Irrigation

Raising water levels using a Terraced Irrigation system appears to be a more flexible option in terms of providing the prescribed water level conditions in the desired areas. This method has been implemented at Wet Moor. The process involves the positioning of penning structures at selected locations to retain upstream water levels to a specified level. The selection of the water levels will be determined by assessment of the mean field level of the hydrological area that the structure(s) serve and the target water level to ground level freeboard previously discussed.

Technically, this is the more complex method by which water levels may be raised but this subtle control of the water environment enables the provision of the desired freeboards in large areas.

Topography, the Selection of the Retained Water Level and Positioning of Structures

The selected location of each structure is critical to the success of retaining water within an area and is dependant upon detailed knowledge of its hydrology. This knowledge, combined with land level information of the area, enables:

i. The positions where structures are required to be identified and

ii. the selection of an appropriate retained water level.

Water Level Management

Clearly with an intensive water level management regime comprising numerous structures, and utilising many of the existing poorly maintained channels in the area there is an increased management and maintenance resource implication. At Wet Moor and all other raised water level schemes constructed by the NRA, the day-to-day operation of the schemes has been accepted by the local drainage boards. These then in turn have delegated this operation to local beneficiaries of the scheme. The NRA maintain an overall management role.

Flood Storage Capacity

The implementation of strategic retention allows the main river channels in the area to be maintained at their existing lower levels. This allows the efficient evacuation of flood waters to continue. Inevitabley, retention of quantities of water in the moorland will occupy a certain amount of flood storage capacity. However, the strategic positioning of evacuation structures for each hydrological area will provide the means by which this water may be evacuated if required.

OVERVIEW OF THE SYSTEM

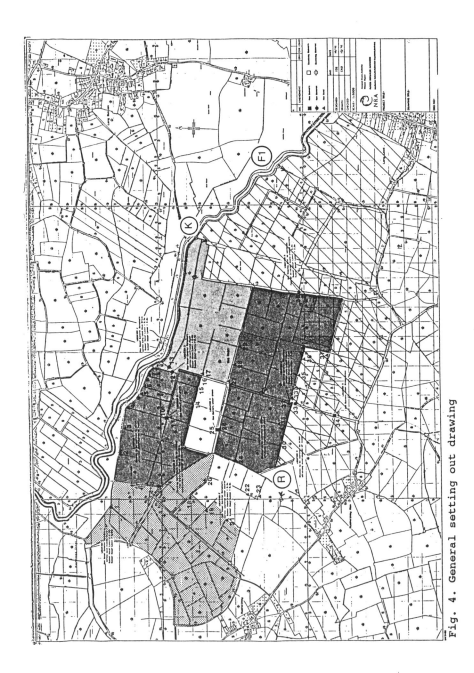

Fig. 4. General setting out drawing

With this more subtle and controllable method of raising water levels the proportion of the flood storage capacity which becomes permanently occupied can be reduced to a minimum. In Wet Moor this reduction of capacity has been calculated at 4% of the total.

Agricultural Activity and the Environment
The use of strategic retention enables the appropriate water level/ground level relationship to be maintained in a large proportion of the area. This ensures that less intensive farming activities can continue to be practised with few of the negative effects from excessively high water levels being experienced

5.2 SCHEME DETAIL

Terraced Irrigation was selected as the means of raising water levels at Wet Moor and indeed all other raised water level schemes implemented under the strategy to date. At Wet Moor this involved the emplacement of:

6 additional water level control structures. (Figure 5)
1.6km of Minor Channel Improvements
30 Earth Dams (Figure 6)
and 10 culverts

to divide the area up into 6 fully isolated hydrological blocks as shown in Figure 4. The area is now operated by feeding water from the main arterial channels at locations F1. K & R to the hydrological blocks with the highest land level elevation. The new water level control structures for these areas are adjusted to retain a seasonal water level appropriate to mean field level of all enclosures in this area. When this water level is achieved water overtops the control structure and cascades into the next hydrological block at a lower level. This process continues through each hydrological block until excess water from the lowest block discharges to the pumped channels as was the case pre-raised water levels. This system has ensured that:

i. Prescribed freeboards may be maintained accurately.

ii. Raised Water Levels are retained only where agreement has been achieved.

iii. Only minimal flood Storage capacity has been lost.

Costs
The capital costs of producing the Wet Moor scheme were £63,000 ie £97/acre. Due to the exceptionally large extent of the site this cost was unusually low. Experience of smaller sites where more structures per unit area have been required has resulted in average costs of between £150 to £250/acre.

Progress
Since the launch of the Strategy in 1992 8 such schemes have been completed in collaboration with English Nature and MAFF. These schemes cover a total area of 1771 acres. Six of these have been directly sponsored by the NRA. A further scheme to produce another 300 acres of raised water levels is currently under construction at Walton Moor and is due for completion at the end of June 1994. Appendix 1 gives details of these areas.

OVERVIEW OF THE SYSTEM

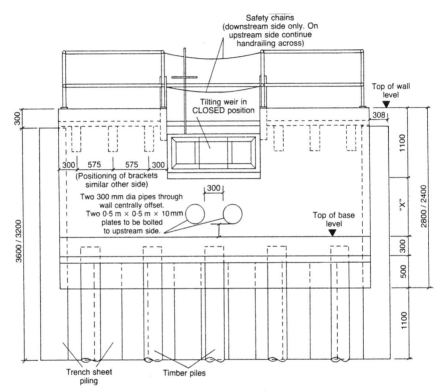

Fig. 5(a). Details of reinforced concrete control structure and tilting mechanism: downstream elevation

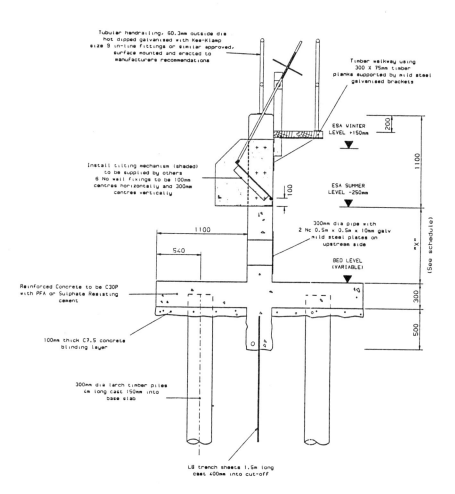

Fig. 5(b). Details of reinforced concrete control structure and tilting mechanism: general arrangement

OVERVIEW OF THE SYSTEM

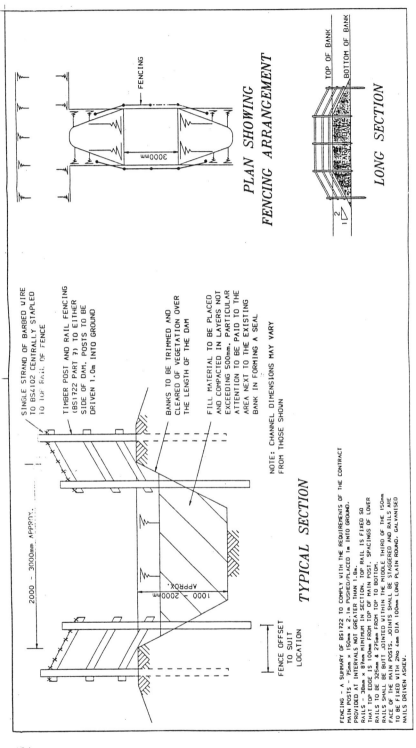

Fig. 6. Typical detail of earth bund and fencing

The Future
    The NRA has made provision for further raised water level schemes in its capital works programme for the next 2 years. This commitment is reviewed annually. A further 10 possible schemes are currently under investigation.
In addition, the NRA is reviewing its water level management on the levels and moors as a whole and will compare these against ideal conditions identified by the various conservation bodies to quantify the gap between current practice and their perception.
MAFF's initiative on Water Level Management Plans then requires the NRA to liaise with all local interests to find an agreed water level management.

APPENDIX 1

SOMERSET LEVELS AND MOORS
RAISED WATER LEVEL AREAS

SUMMARY SHEET

Operational Raised Water Level Areas   (Figure 1.)

Wet Moor	655 Acres
West Moor	127 Acres
Tealham Moor	120 Acres
Southlake Moor	107 Acres
West Sedgemoor	605 Acres
Hay Moor	35 Acres
Town Tree Farm	60 Acres
West Moor Extension	63 Acres
Walton Moor	300 Acres (under construction)

2,072 Acres (838 ha.)

Raised Water Level Schemes Under Development (Figure 1.)

West Sedge Moor	100 Acres
North Moor	200 Acres
East & West Waste	82 Acres
Chilton Moor	147 Acres
Hay Moor Extension	65 Acres
Kings Sedge Moor and Moorlynch	To be confirmed
Town Tree Farm	30 Acres

A further 10 sites where interest in raised water levels has been expressed have been notified to the Authority by MAFF/ADAS.

105

The Rhine-Meuse Delta: ecological impacts of enclosure and prospects for estuary restoration

H. SMIT, National Institute for Coastal and Marine Management,
R. SMITS, Rijkswaterstaat, and H. COOPS, Institute for Inland Water Management and Waste Water Treatment

SYNOPSIS. The Rhine-Meuse Delta was enclosed in 1970 after the completion of the Haringvliet dam. Developments in morphology, water and sediment quality and biology are described. Ecological impacts of enclosure on the Delta, the river Rhine and the North Sea are summarized. Changes in socio-economic needs in the 1980's led to a reconsideration of the area's water management. Three management alternatives are compared and evaluated from an ecological point of view.

INTRODUCTION
Estuaries link rivers to the sea. They form an essential chain between the marine and river ecosystems. This was true for the rivers Rhine and Meuse, until November 1970, when the completion of the Haringvliet dam created a barrier between the rivers Rhine and Meuse and the North Sea. The first aim of the closure was to safeguard the surrounding land against floods. As a part of the Delta works, the dam created a shorter and much safer coastline (ref. 1). The second aim was a better control of the distribution of river water over its main outlets, the Nieuwe Waterweg and Haringvliet. This was needed to reduce salt intrusion into the Nieuwe Waterweg. The third aim was to improve the freshwater supply to the surrounding former islands for agricultural purposes and to provide new and more suitable inlets for drinking water production.
The closure, however, had a dramatic impact on the ecology of the former estuary (ref. 2). Moreover, in the first half of the 1980's, various negative long term effects of the sluices became apparent and were recognized, such as the accumulation of large amounts of contaminated sediments and the disappearance and degradation of former and remaining intertidal areas.

In the second half of the 1980's, integrated water management was introduced, which aimed at taking into account all functions and properties of water systems. Simultaneously, the natural values and potentials of river ecosystems (ref. 3) and coastal wetlands (ref. 4) were becoming increasingly recognized and appreciated in Dutch society. This change was formalized in the National Policy Document on Watermanagement (ref. 5) and the National Nature Policy Document (ref. 6), which aimed at a large scale restoration of nature.

Since the area's ecosystem had become strongly dependent on the management of the Haringvliet sluices, this management had to be reconsidered with the aim of finding a new balance between all interests involved.

This paper aims 1) to summarize the ecological impacts of the Haringvliet sluices on the area and its surroundings; 2) to formulate alternative sluice management strategies and describe their impact on natural and human functions; and 3) to evaluate these strategies from an ecological point of view.

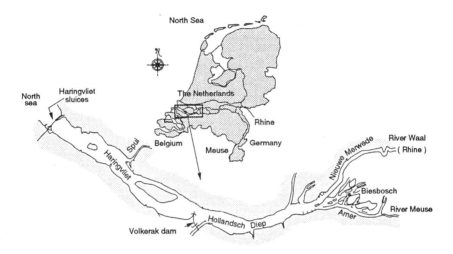

Fig. 1. The enclosed Rhine-Meuse Delta in the Netherlands.

THE (ENCLOSED) RHINE-MEUSE DELTA

The area considered comprises the Haringvliet, Hollandsch Diep, Biesbosch and the rivers Nieuwe Merwede and Amer (Fig. 1). The Haringvliet is the main outlet of the rivers Rhine and Meuse.

The former estuarine ecosystem harboured very high natural values. The vegetation consisted of large stands of reeds and rushes, merging in the east into a large freshwater tidal area, the Biesbosch, with tidal forests and reed marsh areas. Both the invertebrate and fish fauna followed an environmental gradient, with several marine species in the western Haringvliet, brackisch water species in the eastern Haringvliet and Hollandsch Diep, and freshwater species in the Biesbosch, Amer and Nieuwe Merwede. Very high numbers of migrating wader species occurred in the intertidal areas.

The construction of the Volkerak dam (finished in 1969) and Haringvliet dam (finished in November 1970) was needed to convert the estuary into an inland freshwater basin. The following aspects played a major role in the design of the sluices in the Haringvliet dam (the Haringvliet sluices): the discharge of large amounts of river water, the discharge of ice floes during severe frost, the occurrence of waves and currents, bottom stability and the role of the sluices in the distribution of freshwater over the Netherlands. The complex consists of 17 sluices, each about 60 m wide and provided with a pair of steel gates (ref. 2).

Since November 1970, the Haringvliet sluices regulate the water distribution in the basin. At low Rhine discharges ($Q < 1700$ m^3 s^{-1}), the sluices are closed to prevent inland salt intrusion. With increasing Rhine discharge the sluices are gradually opened during low tide on the North Sea. At high Rhine discharges ($Q > 9000$ m^3 s^{-1}) the sluices are completely open.

CHANGES AFTER ENCLOSURE
Morphology.

After the enclosure, channel profiles became oversized and consequently large amounts of sediments were deposited in the channels (ref. 7). These sediments mainly originated from the rivers Rhine and Meuse. The tidal channels of the Nieuwe Merwede filled up first in the early 1970's. In later years, sediment deposited further west as the upstream parts had reached a new equilibrium with flow rates.

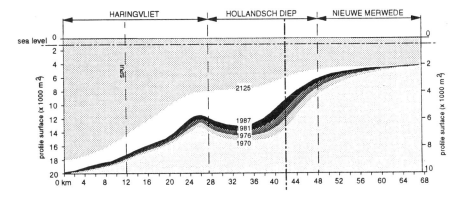

Fig. 2. Surface of cross-sectional profiles in the enclosed Rhine-Meuse Delta between 1970 and 1987, and a prediction for 2125. Surface area of the eastern Hollandsch Diep (to the right of the bold line) refers to the North side, connected to the Nieuwe Merwede. Adapted from ref. 7.

Between 1970 and 1987 cross-sectional profiles of the Amer were reduced by 15-20%, those of the Nieuwe Merwede by 10-15% and those of the eastern Hollandsch Diep by 20% (Fig. 2), coinciding with an average net sedimentation of 0.5-2 m. With the present sluice management, the sedimentation process will continue until a new equilibrium has been established within 1 to 2 centuries.

On the banks erosion by wind waves has changed the gentle slopes of the small remaining areas into steep gradients. The ecologically interesting intertidal zone practically disappeared in most places, except for a few sheltered flats. For example, the vegetation border of wind-exposed intertidal areas in the Haringvliet has receded by about 100 m between 1970 and 1984. During the 1980's most banks have eventually been protected with wave breaks at some distance from the margins of the present terrestrial zone. The protection programme will be finished by 1997. The intertidal gradient has been restored on two flats in the Haringvliet by means of sand nourishment.

Water and sediment quality.
   The above mentioned sedimentation patterns have also influenced the water quality. Parameters related to suspended particles show a decreasing (e.g. total PO4-P, total Lead, chlorophyll-a) or increasing (transparency) trend from the upstream river sections to the Haringvliet dam. Between 1970 and 1990 concentrations of ammonium decreased and of nitrate+nitrite increased, mainly because of the increased rate of purification of domestic waste water. Phosphate showed a decrease since 1986, which is probably related to the increased use of detergents without phosphate. In spite of these changes nutrient concentrations remained high. However, chlorophyll-a concentrations in the Haringvliet were low throughout the period (<20 mg m^{-3}) and transparency relatively high (Secchi disk depth 1-1.5 m).

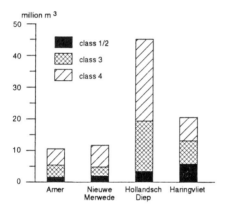

Fig. 3. Amounts and quality of sediments deposited in the Amer, Nieuwe Merwede, Hollandsch Diep and Haringvliet between 1970 and 1989. Quality classes according to ref. 5; class 1: meets the standards of environmental quality (AMK 2000); class 2: moderately polluted; class 3: polluted; class 4: extremely polluted. Classification is done using corrected concentrations of contaminants according to ref 5. Data from Rijkswaterstaat, Zuid-Holland Directorate.

   Sediment quality is mainly determined by the pollution level of the sediments that were deposited after 1970. Sediments deposited in the early 1970's have the poorest quality (pollution class 4), since pollution was then at its peak. The most recent sediments are less polluted (class 2 or 3). Consequently, sediment quality at the surface improves going from the Amer and Nieuwe Merwede to the Hollandsch Diep. However, most of the 90 million m^3 deposited between 1970 and 1989 is polluted (class 3) or extremely polluted (class 4), and is concentrated in the Hollandsch Diep (Fig. 3).

Biotic developments.
   After enclosure, the vegetation has changed drastically. The *Scirpus* (Bulrush) zone gradually deteriorated throughout the basin: between 1970 and 1989, the *Scirpus* stands in the Haringvliet and Hollandsch Diep declined from several hundreds of hectares to less than 1 ha (ref. 8). A similar decline occur-

red in the Biesbosch (ref. 9). Most of the reduction was caused by permanent inundation, accompanied by erosion and heavy grazing by geese. After the disappearance of *Scirpus* stands, wave attack affected the *Phragmites australis* (Reed) vegetation. Locally, this led to the formation of steep shoreline cliffs without emergent vegetation at the water's edge.

In the higher parts, the absence of frequent inundation led to oxygen penetration into the bottom and hence to an increased availability of nutrients. These areas were soon invaded by terrestrial ruderal vegetation species such as *Urtica dioica* (Common nettle), *Epilobium hirsutum* (Great hairy willow-herb) and *Solidago gigantea* (Golden rod).

The macro-invertebrate fauna since the closure has been rather poor in species with low biomass, and has been generally dominated by a few tolerant tubificid worm species (ref. 10). Exceptions are the western Hollandsch Diep and eastern Haringvliet, where high *Dreissena polymorpha* (Zebra mussel) biomasses occur. The mussels are an important food source for e.g. *Aythya fuligula* (Tufted duck).

After the enclosure, all marine and migrating fish species had soon disappeared, except for the katadromous *Platichthys flesus* (Flounder) and small numbers of *Osmerus eperlanus* (Smelt). A rapid colonization of common freshwater species such as *Gymnocephalus cernuus* (Ruffe), *Perca fluviatilis* (Perch), *Rutilus rutilus* (Roach) and *Abramis brama* (Bream) followed. From 1976 onwards, *Stizostedion lucioperca* (Pike-perch) appeared; its importance gradually increased in the coarse of the following decade. After 1980, Bream has become dominant, replacing Roach (ref. 11).

Water bird monitoring data collected in winter show, that the Haringvliet and Hollandsch Diep area is still a wetland of international importance. Maximum numbers of three bird species exceeded the 1% standard in five consecutive years between October 1988 and March 1993: *Brantha leucopsis* (Barnacle goose, range of maxima: 20-34%), *Anser anser* (Greylag goose: 2.7-6.1%) and *Anas penelope* (Wigeon: 1.4-3.3%). However, these species mainly live on the surrounding grasslands, ecosystems which function largely independent of the aquatic ecosystem. During the last four years, Tufted duck (1.0-2.0%) also exceeded the 1% standard.

ECOLOGICAL IMPACTS OF PRESENT SLUICE MANAGEMENT

This section summarizes the main ecological impacts of the present sluice management, both on the system itself and on the connected rivers and the North Sea.

1. *Disturbed silt balance*. Silt, including organic material, has accumulated in the channels and river beds of the Nieuwe Merwede and the eastern Hollandsch Diep, where it has become for the greater part unavailable to the ecosystems both of the area itself and of the North Sea; in the area itself it remains unused, while smaller quantities now reach the North Sea. The former tidal flats have been deprived of building materials and of their main food source, the organic fraction of the sedimenting silt.

2. *Accumulation of contaminated sediments.* Most of the sediment accumulated in the area is severely polluted. This produces a hazardous situation for both man and nature. Several toxic compounds, such as chlorobiphenyls, reach high levels in the sediment and approach the (high) consumption standard in Eel (*Anguilla anguilla*) (ref. 12), which is still commercially fished in the area.

Negative impacts on various components of the ecosystem have also been revealed: the reproduction of *Phalacrocorax carbo sinensis* (Cormorant) (ref. 13) and Tufted ducks (ref. 14) was negatively affected; there was a high incidence of malformations of the head capsules of chironomid larvae (ref. 15) and densities of chironomids were low (ref. 10).

The costs of removing the contaminated sediments are very high: 125 million US $ to remove all class 4 sediments (the category of most polluted sediments) and about 400 million US $ to remove all class 3 and 4 sediments. Thanks to the sluices, however, the contaminated sediments have not spread into the North Sea, which facilitates removal.

3. *Disappearance of intertidal areas.* After the enclosure most of the intertidal areas became either permanently submerged or exposed. The brackish tidal areas have disappeared completely. The remaining freshwater intertidal areas declined further, due to the erosive power of wind waves. Consequently, only a few hundred hectares of freshwater tidal area remain 20 years after enclosure. Both the freshwater and brackish intertidal areas harbour biotopes of international importance (e.g. for migrating water bird species) and their decline or disappearance is regarded as a serious ecological loss.

4. *Increased flat formation in the outer delta.* In the outer Haringvliet Delta, on the North Sea side, channels have filled up and new intertidal sandbanks have developed, as a result of the decreased tidal currents. These areas also have high natural values, since they harbour large numbers of several water bird species, have a nursery function for fish and house a characteristic macro-invertebrate fauna (ref. 16).

5. *Disappearance of nursery function for fish.* The former estuary had a nursery function for several marine fish species such as the Sole *(Solea solea)* and Plaice (*Pleuronectes platessa*) (ref. 17). This nursery function has disappeared and is now restricted to the outer delta.

6. *Disturbance of fish migration.* Formerly common anadromous fish species have disappeared from the rivers Rhine and Meuse for several reasons. One was the closure of most river outlets. The countries along the river Rhine adopted the Rhine Action Programme in 1987. One of the objectives is the return of anadromous fish species, such as *Salmo salar* (Salmon). Numerous measures are presently being taken in the river basin to prepare it for the return of this fish species. The present functioning of the sluices, however, forms an important obstacle (ref. 18). When the sluices discharge, current velocities are too high to allow any significant upstream migration. At low Rhine discharges the sluices are closed and no passage is possible at all.

OVERVIEW OF THE SYSTEM

7. *Hampered mixing of river and sea water.* Before closure, an intensive mixing of river and sea water occurred in the estuary. The Delta project has diminished the rate of mixture dramatically. Since the enclosure of the Haringvliet, a larger proportion of the river water flows through the Nieuwe Waterweg, a small and deep channel, where circumstances for mixing are unfavourable. Consequently, fresh water is now being injected into the North Sea, where most of the mixing with sea water takes place. Laane and coauthors (ref. 19) showed that the average salinity measured at a group of nearshore stations (< 10 km from the coastline) along the Dutch coast has dropped since 1970, while nutrient concentrations e.g. of dissolved inorganic nitrogen (D.I.N.) have increased (Fig. 4). Consequently, more nutrient rich river water is transported along the Dutch coast to the Wadden Sea. These results indicate that the Delta project, of which the Haringvliet sluices constitute an important component, may have contributed to eutrophication problems along the Dutch coast.

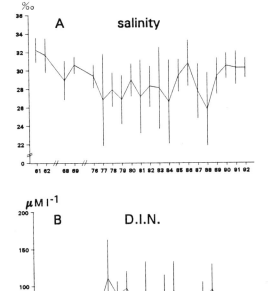

Fig. 4. Annual variation in salinity (‰) and dissolved inorganic nitrogen (D.I.N., $\mu M\ l^{-1}$) during the winter (January - March) in the North Sea from 1961 to 1992. Data are averages (± 1 SD) of a group of localities along the Dutch coast (< 10 km offshore). Courtesy Laane et al. (ref. 19).

8. *In conclusion.* The present management of the Haringvliet sluices has an impact on the Rhine ecosystem up to central Germany and on the North Sea ecosystem along the Dutch coast, as well as a major impact on the ecosystem of the Rhine-Meuse Delta itself.

ALTERNATIVES TO THE PRESENT SLUICE MANAGEMENT
Recent policy documents.

Recent years have seen the publication of a number of policy documents in the Netherlands, in which emphasis shifted to ecology in physical planning and water management, and in which a new approach to nature conservation is presented. This new approach was implemented in the *Third National Water Policy document* (ref. 5). In this document, the Dutch government promised a new policy for the basin, including a reconsideration of the management of the Haringvliet sluices. This new policy was worked out in the *Integrated Policy Document on the Haringvliet, Hollandsch Diep and Biesbosch*, referred to below as IPDH. This document provides an inventory of properties and functions (ref. 20), an analysis of alternatives and selection procedures (ref. 21) and a detailed description of the selected policy for the management of the Haringvliet sluices and the removal of contaminated sediments (ref. 22).

Removal and storage of contaminated sediments.

Studies for the IPDH have shown, that new sluice management would rapidly lead to the remobilization of contaminated sediment and its transport to the North Sea, which was widely considered unacceptable. Therefore, the most contaminated sediments would have to be removed first before a new sluice management could be implemented. The possibilities for removal of contaminated sediments are now under study. The removal operation will probably start in 1997 and last about 20 years. The thick layers in the eastern part of the basin will be removed using a cutter suction dredge. In shallow areas, other methods such as the grab will be used. In the Haringvliet, the water injection dredging method can be used. By injecting water, the silty top layer is made fluid and the fluid mud will flow to nearby channels, from where it can be removed. Recent field experiments show promising results. A large scale sludge dump which could store about 30 million m^3 of contaminated sediment has been projected in the western Hollandsch Diep. The dumping site is oval shaped (length*width = 1.2*0.9 km) and the dike is elevated only 3.5 m above mean water level for landscape reasons. The site will be excavated to a depth of 50 m below mean water level. The non polluted suitable sand will be used for infrastructural projects, while the other non polluted sediments (at least 10 million m^3) will be used to restore or develop several hundreds of hectares of intertidal areas. The total costs of the dumping site project is estimated to be about 200 million US $.

Management alternatives.

In the IPDH, various sluice management alternatives were studied, varying from continuing the present management to opening the sluices and using them as a storm surge barrier only. Three alternatives are presented in this paper, which represent the full range of variants: 1) the present situation,

OVERVIEW OF THE SYSTEM

Table 1. Characteristics of three management alternatives of the Haringvliet sluices. Surface areas of biotopes denote areas which have the potential for development as such. MHW: mean high water level; MLW: mean low water level; mfl: million guilders (1 guilder = 0.5 US $).

	sluice	management	alternative
	present management	preferred by IPDH	storm surge barrier
Physical characteristics	HV0	HV2	HV4
mean tidal range (m)	0.3	0.55-0.95	1-1.4
average period of salt water inlet (months)	0	6	12
salt intrusion up to	sluices	Spui	Holl. Diep
Water and sediment quality in the area	poor	intermediate	good
mobility of contaminated sediments	little	limited	considerable
Important biotopes (km^2)			
brackish marshes	0	3.5	4
reed/weed vegetation and grassland from MHW to dike	43	46.5	39
helophyte stands (reeds and rushes)	16.5	15.5	18
softwood floodplain forest (Salix)	5	8	5
hardwood floodplain forest	30	35	25
intertidal area (MLW-MHW)	10.5	19.5	32.5
submerged waterplant vegetation (MLW-75 cm - MLW)	16.5	7	8.5
Functions/costs			
agricultural compensation costs (mfl)	0	52-176	909
drinking water compensation costs	0	18	103-123
navigational depth	unchanged	some reduction	reduction at sluices
conditions for fisheries (mfl)	poor	intermediate	good
recreation, adaptation costs (mfl)	0	4.2	6.8
swimming water quality	poor	intermediate	good
general attractiveness	unchanged	slight increase	increase
adaptation costs Haringvliet sluices (mfl)	0	18	1

referred to below as HV0; 2) the alternative preferred in the IPDH, referred to as HV2, and 3) the storm surge barrier alternative, referred to as HV4.

Characteristics of the three alternatives and their impacts on the ecology and various anthropogenic functions have been summarized in Table 1. Details are given in ref. 21. From an ecological point of view, HV4 is very attractive, since it leads to the creation of large areas of very valuable biotopes. The costs, however, are estimated to be very high. In the IPDH, HV2 has been chosen as the preferred alternative, since it combines some profit for nature with acceptable costs.

A PLEA FOR THE RESTORATION OF ESTUARINE PROCESSES
Significance of estuary restoration.

Since the Rhine-Meuse estuary is the connection between the basins of the rivers Rhine and Meuse and the North Sea, the management of this estuary is highly relevant to the rehabilitation programmes of these large rivers (R.A.P.: Rhine Action Programme) and the North Sea (N.A.P.: North Sea Action Programme).

The severe degradation of the estuarine ecosystem which occurred over the last 20 years is hampering the execution of both action programmes. Good possibilities for the passage of migratory fish species are necessary to meet one of the main R.A.P. objectives: the return of indigenous fish species such as the Salmon.

Better mixing of fresh and salt water in the estuary will help to solve eutrophication problems in the North Sea, a major N.A.P. objective. Moreover, an estuary can be considered as a large scale nutrient filter (ref. 23). Both chemical and biological processes contribute to this filter function.

The intertidal areas in the entire Delta area have a very important function as a refuel site for migrating birds and as a wintering site for birds breeding in northern Europe.

If the basin is to be managed as an estuary, the HV4 alternative seems the only valid option. Only HV4 offers a considerable exchange of tidal volumes, a large intertidal area and a constant tidal rhythm, factors which are essential for the existence of estuarine communities.

In contrast, the HV2 alternative is less attractive than it would appear from the biotope surfaces created. HV2 is not merely an intermediate between HV0 and HV4; regular tidal movement is completely lacking. It would mean that the duration of sluice closure is far too long for most intertidal species to survive. In this period fish migration is blocked and in addition, HV2 offers less mixture of fresh and salt water.

In the IPDH, external impacts of the sluices have played only a minor role. On the other hand, the total costs of nutrient removal -part of the North Sea Action programme- are high.

The path to restoration.

If the estuary is to be restored, this should be done as soon as possible, since irreversible processes including the geomorphological changes and the extinction of relict species are continuing. On the other hand, the problem of polluted sediments should be solved first, so as to prevent their spread. The ideal restoration path (Fig. 5) would involve simultaneous approaches to both the habitat and the pollution problems. Investing in the one and omitting the other will lead to suboptimal returns on investments. Hence, the polluted sediment removal programme (sanitation) should be followed as soon as possible by a gradual further opening of the sluices, ultimately ending with a complete opening (HV4). This simultaneous tackling of both problems can be regarded as the pathway to sustainable development; it develops the area in a sustainable way by working with the natural processes and not against them.

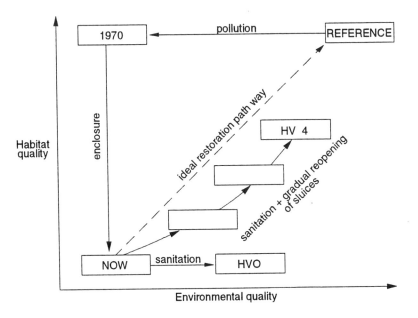

Fig. 5. Conceptual model of the path of ecosystem degradation including pollution and habitat destruction, and of possible restoration including the removal of polluted sediments (sanitation) and a stepwise reopening of the sluices.

CONCLUSION

In the past, man has learned to control the estuarine system. This study has shown that overcontrol may lead to undesirable ecosystem developments. However, the same tool (the Haringvliet sluices) can be used to manage the basin as an estuary. It is a challenge to restore the estuary without increasing the threat of a catastrophic inundation, which prompted man to enclose it nearly 25 years ago.

ACKNOWLEDGEMENTS

The authors would like to thank Prof. Dr. G. van der Velde, Prof. Dr. J. Pethic, ir. L. Bijlsma, Dr. G.T.M. van Eck, and Drs. R. van Otterloo for their valuable comments on the manuscript, and mrs. S van Pamelen and mr. H. van Reeken for technical assistance.

REFERENCES

1. SAEIJS H.L.F. Changing estuaries: a review and new strategy for management and design in coastal engineering. Rijkswaterstaat, The Hague, 1982, Government Publishing Office; ISBN 9012039215.
2. FERGUSON H.A. and W.J. WOLFF. The Haringvliet-project: the development of the Rhine-Meuse estuary from tidal inlet to stagnant freshwater lake. Wat. Sci. Tech., 1983, vol. 16, 11-22.
3. DE BRUIN D., D. HAMHUIS, L. VAN NIEUWENHUIZEN, W. OVERMARS and F. VERA. Plan Ooievaar. Gelderse Milieufederatie, Arnhem, 1987.
4. MINISTRY OF HOUSING, SPATIAL PLANNING AND THE ENVIRONMENT. "Planologische Kernbeslissing Waddenzee": parts 1, 2 and 3; The Hague, 1993, ISBN 9039903956 [in Dutch].
5. MINISTRY OF TRANSPORT AND PUBLIC WORKS. Water in the Netherlands: a time for action. Summary of the Third National Water Policy Document, The Hague, 1989 [in English].
6. MINISTRY OF AGRICULTURE, NATURE MANAGEMENT AND FISHERIES. Nature policy document: governmental decision, The Hague, 1990, ISBN 9012069017 [in Dutch].
7. VAN BERGHEM J.W., M.A. DAMOISEAUX and P.F. VAN DREUMEL. Geomorphological mapping of the Haringvliet, Hollandsch Diep, Nieuwe Merwede and Amer. Report and 6 maps; Rijkswaterstaat, Zuid Holland Directory, Rotterdam, 1992 [in Dutch].
8. SMIT H. and H. COOPS. Ecological, economic and social aspects of natural and man-made bulrush (*Scirpus lacustris* L.) wetlands in The Netherlands. Landscape and Urban Planning, 1990, vol. 20, 33-40.
9. COOPS H. Historical changes in foreland marshes in the Northern Delta area and the delta of the River IJssel, The Netherlands. Report 92.030 of the Institute for Inland Water Management and Waste Water Treatment/RIZA, Lelystad, 1992 [in Dutch].
10. SMIT H., H.C. REINHOLD-DUDOK VAN HEEL and S.M. WIERSMA. Macrozoobenthic densities, biomasses and species composition in relationship to environmental parameters in the enclosed Rhine-Meuse Delta. Submitted to Neth. J. Aquat. Ecol.
11. WIEGERINCK J.A.M. and M.J. HEESEN. Fisheries observations in the Haringvliet and Hollandsch Diep during 1976 through 1986. Fisheries Directorate, Report 31, The Hague, 1988 [in Dutch].
12. HENDRIKS A.J. and H. PIETERS. Monitoring concentrations of microcontaminants in aquatic organisms in the Rhine Delta: a comparison with reference values. Chemosphere, 1993, vol. 26, 817-836.
13. DIRKSEN S., T.J. BOUDEWIJN, L.K. SLAGER, R.G. MES, M.J.M. SCHAIK and P. DE VOOGT. Reduced breeding success of Cormorants (*Phalacrocorax carbo sinensis*) in relation to persistent organochlorine pollution

of aquatic habitats in the Netherlands. Environ. Poll., accepted.
14. DE KOCK W. Chr. and C.T. BOWMER. Bioaccumulation, biological effects, and food chain transfer of contaminants in the Zebra mussel (*Dreissena polymorpha*) In: Zebra mussels: biology, impacts, and control, T.F. Nalepa & D. Schloesser (eds.): CRC Press, Florida, ISBN 0873716965, 1992, p 503-533.
15. VAN URK G. and F.C.M. KERKUM. Misvormingen bij muggelarven uit Nederlandse oppervlaktewateren. H_2O, 1986, vol. 19, 624-627 [with English summary].
16. RIJKSWATERSTAAT. The Voordelta: a watersystem in change. Report of Tidal Waters Division, The Hague, 1989 [in Dutch].
17. VAAS K.F. The fish fauna of the estuaries of the rivers Rhine and Meuse. Biologisch Jaarboek, Dodonaea, 1968, vol. 36, 115-128 [in Dutch].
18. VAN DIJK G.M. and E.C.L. MARTEIJN (eds.). Ecological rehabilitation of the river Rhine. E.H.R report 50, Institute for Inland Water Management and Waste Water Treatment, Lelystad, 1993.
19. LAANE R., R. WILSON, R. RIEGMAN, P.A.L. VAN DER MEYDEN and G. GROENEVELD. Dissolved inorganic nitrogen and phosphate in the Dutch coastal zone of the North Sea, the North Sea and in the Rhine during 1961-1992: concentrations, ratios and trends. Manuscript, to be submitted to Neth. J. Sea Res.
20. RIJKSWATERSTAAT. Integral Policy Document Haringvliet, Hollandsch Diep, Biesbosch. Report of the inventory working group, phase I. Zuid-Holland Directorate, Rotterdam, 1990 [in Dutch].
21. RIJKSWATERSTAAT. Integral Policy Document Haringvliet, Hollandsch Diep, Biesbosch. Report of phase 2, analysis and selection. Main report, summary and partial reports A to L, Zuid-Holland Directorate, Rotterdam, 1991 [in Dutch].
22. RIJKSWATERSTAAT. Integral Policy Document Haringvliet, Hollandsch Diep, Biesbosch: design, final report. Zuid-Holland Directorate, Rotterdam, 1994 [in Dutch].
23. KENNEDY V.S. (ed.). The estuary as a filter. Academic Press, New York, 1984, ISBN 0124050700.

Numerical modelling of hydrodynamic and water quality processes in an enclosed tidal wetland

R. A. FALCONER, Professor of Water Engineering, and S. Q. LIU, Visiting Fellow, University of Bradford

SYNOPSIS. Details are given of the application and refinements of a mathematical model for predicting flow, water quality indicator and sediment transport processes within a large semi-enclosed coastal wetland basin. Particular emphasis has been focused on the numerical representation of the flooding and drying of the tidal wetlands and, in particular, the representation of the advective transport of the solutes and the significance of the decay rates. The model has generally produced accurate predictions of the hydrodynamic and water quality indicator processes, in comparison with field data provided by the local region of the National Rivers Authority.

INTRODUCTION
1. The paper outlines a long term research programme which has been established to improve on the accuracy of a numerical model for predicting the hydrodynamic, water quality constituent and sediment transport processes in a semi-enclosed coastal wetland basin, namely Poole Harbour, in Dorset, UK. Poole Harbour (see Fig.1) is one of the largest natural harbours in Europe, with a maximum perimeter of approximately 112 miles, and is connected to the English Channel by a narrow entrance which is only approximately 300m wide. The harbour is a site of special scientific interest (i.e. a SSSI). Extensive areas of salt marsh and phragmites reed exist within the basin, with these areas being of botanical and ornithological interest which support nationally rare populations of plants and birds. The estuary also provides valuable breeding and nursery areas for several species of fish, including bass, pollack and mullet. The harbour is generally shallow and the tidal range relatively small, i.e. typically 2m. Within Poole Harbour the tidal curve has a distinct characteristic in that a second high tide occurs approximately 3 hours after the first. At low tide, a variety of wildlife abounds on the exposed shores, particularly in the undeveloped areas to the south and west of the harbour. In contrast, the town of Poole - with its quay and commercial port - form part of an expanding and important trading centre for shipping links with Europe and the ecological interest of the basin could be put at risk from severe pollution - particularly in view of the potential for poor flushing with the relatively narrow entrance.

2. With these considerations in mind, particular emphasis has been focused in this study on the numerical representation of the complex hydrodynamic processes associated with tidal flooding and drying of the wetlands and, secondly, the numerical treatment of high concentration gradients occurring in the region of the sewage treatment works at Keysworth, Lytchett Minster and, in particular, Poole. Such gradients frequently arise in the vicinity of effluent discharge outfalls, with the main problem of traditional numerical models being that grid scale oscillations, or undershoot and overshoot, frequently occur in the numerical solution (ref.1). These oscillations can lead to unrealistic numerical predictions of the water quality constituent concentration distributions and can only be overcome by including undesirable artificial diffusion. As an alternative, higher order

Wetland management. Thomas Telford, London, 1994

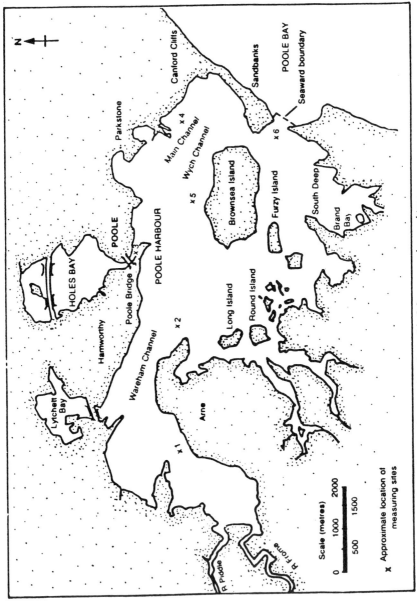

Fig. 1. Map of Poole Harbour showing field measuring sites

accurate numerical schemes can be used to minimise these oscillations, with a minimum amount of artificial diffusion then being required. In the current study a modified form of the ULTIMATE QUICKEST scheme (ref.2) has been applied to Poole Harbour, with a range of water quality indicators being included in the refined model.

3. The study also provides an extension of an earlier study, undertaken for Wessex Water plc, where the main interest at the time was to ascertain the influence of the levels of nitrate input from Poole Sewage Works on the corresponding concentrations across the basin, particularly in view of the growth of green seaweeds Ulva and Enteromorpha in the region of the Main Channel and Wych Channel (Fig.1), and to the north east of Brownsea Island (ref.3). The nitrate inputs of particular interest in the previous study were the concentrations resulting from inputs of both total oxidised nitrogen (TON) and ammoniacal nitrogen (NH_3) from the two river inputs (i.e. the rivers Piddle and Frome) and the three sewage treatment works (i.e. Poole, Keysworth and Lytchett Minster). The negative concentrations were completely eliminated in the study by using the ULTIMATE QUICKEST scheme.

GOVERNING MODEL EQUATIONS

4. The general differential equations for the hydrodynamic model were based on the depth integration of the Navier-Stokes equations, including the effects of local and advective accelerations, the earth's rotation, bottom friction, wind shear and turbulence. The corresponding equations of mass and momentum conservation in the x and y co-ordinate directions, on a horizontal plane, can be shown to be of the following form (ref.4):-

$$\frac{\partial \zeta}{\partial t} + \frac{\partial q_x}{\partial x} + \frac{\partial q_y}{\partial y} = 0 \qquad (1)$$

$$\frac{\partial q_x}{\partial t} + \beta \left[\frac{\partial U q_x}{\partial x} + \frac{\partial V q_x}{\partial y} \right] - f_c q_y + gH \frac{\partial \zeta}{\partial x} + \frac{1}{\rho}\left[\tau_{bx} - \tau_{sx} \right] - \upsilon_t \left[\frac{\partial^2 q_x}{\partial x^2} + \frac{\partial^2 q_x}{\partial y^2} \right] = 0 \qquad (2)$$

$$\frac{\partial q_y}{\partial t} + \beta \left[\frac{\partial U q_y}{\partial x} + \frac{\partial V q_y}{\partial y} \right] + f_c q_x + gH \frac{\partial \zeta}{\partial y} + \frac{1}{\rho}\left[\tau_{by} - \tau_{sy} \right] - \upsilon_t \left[\frac{\partial^2 q_y}{\partial x^2} + \frac{\partial^2 q_y}{\partial y^2} \right] = 0 \qquad (3)$$

where ζ = water surface elevation above (positive) datum, q_x, q_y = depth integrated velocity components in x,y directions, t = time, β = momentum correction factor for non-uniformity of vertical velocity profile, f_c = Coriolis parameter for earth's rotation, g = gravitational acceleration, H = total water column depth, ρ = fluid density, τ_{bx}, τ_{by} = bed shear stress components in x,y directions, τ_{sx}, τ_{sy} = surface wind shear stress components, and υ_t = depth averaged turbulent eddy viscosity.

5. The bed shear stress can be represented in the form of a quadratic friction law, based on the relationship for steady uniform open channel flow, and for the x-direction can be written as:-

$$\tau_{xb} = \rho \frac{f U V_s}{2} \qquad (4)$$

where U = depth averaged velocity in the x-direction, V_s = depth averaged fluid speed and f = Darcy-Weisbach resistance coefficient. For modelling shallow flows over wetlands, Reynolds number effects can become significant and the Darcy-Weisbach friction factor f was evaluated in the model using the Colebrook-White equation (ref.5) given as:-

$$\frac{1}{\sqrt{f}} = -4\log_{10}\left[\frac{k_s}{12H} + \frac{2.5}{R_e \sqrt{f}}\right] \quad (5)$$

where k_s = Nikuradse equivalent sand grain roughness and R_e = Reynolds number (= 4 V_s H/υ - where υ = kinematic laminar viscosity).

6. Likewise, the surface shear stress due to wind action can also be represented in a quadratic law relationship, given for the x-direction as:-

$$\tau_{sx} = C_f \rho_a W_x W_s \quad (6)$$

where C_f = air-water resistance coefficient, formulated in a piece-wise manner as proposed by Wu (ref.6), ρ_a = air density, W_x = wind velocity in x-direction and W_s = wind speed. In addition to incorporating the wind action in the form of a surface shear stress, research by Falconer and Chen (ref.7) showed that the use of a second order parabolic vertical velocity profile, and its impact on the momentum correction factor β, considerably improved on the representation of this complex hydrodynamic process.

7. Finally, in the absence of field data, the depth averaged eddy viscosity υ_t was evaluated by assuming a logarithmic velocity profile (i.e. assuming bed generated turbulence to be the dominant mechanism) and a typical field measured coefficient reported by Fischer et al (ref.8). Hence, the formulation for υ_t was given by:-

$$\upsilon_t = 1.2 \, U_* \, H \quad (7)$$

where U_* = shear velocity (= $\sqrt{\tau_b / \rho}$, where τ_b = bed shear stress).

8. For the various water quality indicators the general form of the three-dimensional advective-diffusion equation was first integrated over the depth (ref.9) to give:-

$$\frac{\partial \phi H}{\partial t} + \frac{\partial \phi q_x}{\partial x} + \frac{\partial \phi q_y}{\partial y} - \frac{\partial}{\partial x}\left[HD_{xx}\frac{\partial \phi}{\partial x} + HD_{xy}\frac{\partial \phi}{\partial y}\right] - \frac{\partial}{\partial y}\left[HD_{yx}\frac{\partial \phi}{\partial x} + HD_{yy}\frac{\partial \phi}{\partial y}\right]$$

$$- H\left[\phi_i + \phi_d + \phi_k\right] = 0 \quad (8)$$

where ϕ = water quality indicator concentration, D_{xx}, D_{xy}, D_{yx}, D_{yy} = depth averaged longitudinal dispersion and turbulent diffusion coefficients in x,y directions, ϕ_i = source or sink input, ϕ_d = decay or growth rate for the water quality indicator and ϕ_k = total kinetic transformation rate.

9. For the dispersion-diffusion coefficients the terms were expressed in a similar manner to the eddy viscosity, i.e. assuming a logarithmic velocity distribution (ref.9), and given for the x-direction as:-

$$D_{xx} = \frac{\left(k_\ell U^2 + k_t V^2\right) H \sqrt{f}}{\sqrt{2} \, V_s} + D_w \quad (9)$$

where k_ℓ = longitudinal dispersion constant (= 13.0 - based on typical field data in ref.8), k_t = turbulent diffusion constant (= 1.2 - based on typical field data) and D_w = wind induced dispersion-diffusion coefficient.

10. For the individual water quality indicator equations, the corresponding representations included in the model were based on the formulations used in the USA EPA QUAL-11 model (ref.10). Thus, for the main water quality indicators considered in this study, namely the nitrogen cycle, the corresponding differential equations used in the study were as follows:-

Ammonia Nitrogen (N_1):-

$$\frac{D N_1 H}{Dt} - DSPDF = \beta_3 N_4 H - \beta_1 N_1 H + \sigma_3 \tag{10}$$

Nitrite Nitrogen (N_2):-

$$\frac{D N_2 H}{Dt} - DSPDF = \beta_1 N_1 H - \beta_2 N_2 H \tag{11}$$

Nitrate Nitrogen (N_3):-

$$\frac{D N_3 H}{Dt} - DSPDF = \beta_2 N_2 H \tag{12}$$

Organic Nitrogen (N_4):-

$$\frac{D N_4 H}{Dt} - DSPDF = -\beta_3 N_4 H - \sigma_4 N_4 H \tag{13}$$

where D/Dt = total derivative, DSPDF = dispersion-diffusion terms, β_1 = rate constant for biological oxidation of ammonia nitrogen, β_2 = rate constant for oxidation of nitrite nitrogen, β_3 = organic nitrogen hydrolysis rate, σ_3 = benthos source rate for ammonia nitrogen, and rate coefficient for organic nitrogen settling, σ_4 = rate coefficient for organic nitrogen settling.

MODEL DETAILS AND APPLICATION

11. The governing general differential equations (1) to (3) and (8) were solved in the mathematical model using the finite difference technique. This method requires that a regular mesh of square grids be set up over the region of interest and the governing differential equations re-formatted using Taylor's mathematical series. The corresponding finite difference equations are solved for each grid square, with the conservation equations of mass and momentum in the x and y directions giving the water surface elevation ζ and the depth averaged velocity components U, V, and the advective-diffusion equation for each water quality constituent giving the corresponding concentration distributions. For the current study the grid size was set to 150m, leading to a mesh of 65 x 58 grid squares. The local depth below datum was included at the centre of the side of each grid square, with the depths below datum being obtained from the Admiralty Chart and Poole Harbour Commissioners, and the bed roughness height k_s being specified at the centre of each square in the form of a constant.

12. The finite difference equations were then solved implicitly at each time step, using the Alternating Direction Implicit (ADI) formulation and the method of Gauss elimination and back substitution. The time step was set to 72s with the maximum Courant number (i.e. $\Delta t \sqrt{g H} / \Delta x$, where Δt = time step and Δx = grid size) being set to approximately 8 for accuracy limitations (ref.1). The model was always started from rest (i.e. zero velocity and a constant water elevation everywhere), with the open boundary

water elevations being driven by tidal data provided by Poole Harbour Commissioners and with the river and sewage discharges and water quality indicator inputs being provided by Wessex Water plc and the National Rivers Authority (Wessex Region). Field measurements of various water quality indicators were taken at 12 sites across the basin and direct comparisons were made with the numerical model predictions at these sites.

13. In the first part of the model developments relating to modelling the flow over the wetlands in Poole Harbour, emphasis was focused firstly on refining the numerically complex hydrodynamic processes of flooding and drying, with the plan-form area of Poole Harbour varying considerably (typically by up to 50%) during the tidal cycle. The refined treatment of the flooding and drying process was based on a scheme first tested for idealised geometries including, in particular, uniformly sloping and combined sloping and horizontal bed sections (ref.7). The refined scheme gave more robust and realistic solutions than several other documented schemes, with the final scheme being the best of a range of variations considered. When the refined scheme was applied to tidal predictions in Poole Harbour and Holes Bay, and compared with an earlier scheme by Falconer and Owens(ref.11) and a scheme by Stelling et al (ref.12), the new scheme gave more accurate and smoother predictions of the field measured water elevations at Poole Bridge and with the results being similar to those illustrated by Falconer and Chen (ref.7). The new scheme also predicted the moving boundary location more consistently with the basin's bathymetry and this was partly attributed to the inclusion of Reynolds number effects in the representation of the bed shear stress term.

14. In the second part of the refinements to the model for the Poole Harbour study, various water quality indicators were included in the model and were refined, calibrated and verified against extensive field data acquired specifically for this project by the National Rivers Authority (Wessex Region). One of the main research developments relating to this part of the project was the need to investigate the influence of varying decay rates and, in particular, to refine the numerical treatment of high concentration gradients to predict the solute concentration distributions more accurately and to eliminate the occurrence of negative concentration predictions (or undershoot) associated with such gradients. In a detailed study of 36 finite difference schemes for modelling severe concentration gradients (ref.2), the ULTIMATE QUICKEST scheme was found to be computational efficient and highly accurate, and gave rise to no overshoot or undershoot. This scheme was then incorporated into the Poole Harbour model and was used to predict the full range of nitrogen constituents. The scheme had to be refined to incorporate the extensive flooding and drying within the basin and was generally found to give good agreement with the field measured data

15. Typical examples of the corresponding numerical model predictions of the ammonia nitrogen concentration distributions across Poole Harbour are shown in Figs. 2 and 3, at low water level during the fifteenth tidal cycle, both with and without input from Poole Sewage Treatment Works respectively. Velocities are shown only at every other grid point, with the results obtained using the ULTIMATE QUICKEST scheme showing a sharp plume and peak concentration levels falling off rapidly within a short distance of the outfall sites. The comparisons between the predicted and the field measured results at three typical sites are shown in Fig.4, and confirm the good agreement between both sets of results. The results also confirm the need for further research into the influence of the decay rate, since there is no consistency in Fig.4 between the field data and an increase or decrease in the decay rate. Finally, the simulations undertaken both with and without input from Poole Sewage Treatment Works suggested that the effluent input from the sewage works had some effect on the nitrate levels to the north east of Brownsea Island. However, in contrast the nitrate levels in Holes Bay were significantly reduced with the exclusion of input from Poole Sewage Treatment Works.

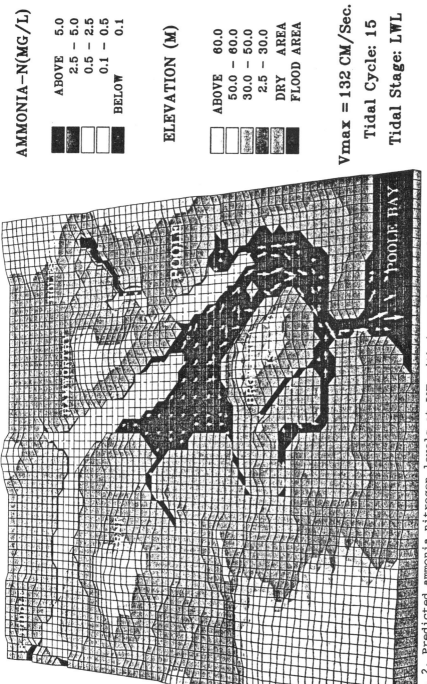

Fig. 2. Predicted ammonia nitrogen levels at LWL with input from Poole STW

HYDRAULIC PROCESSES

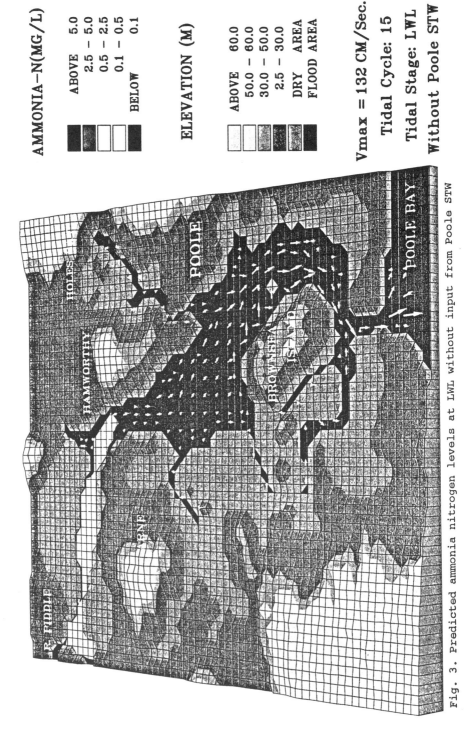

Fig. 3. Predicted ammonia nitrogen levels at LWL without input from Poole STW

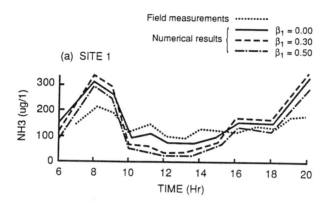

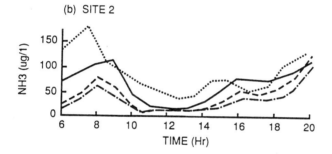

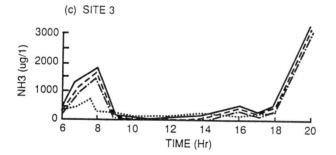

Fig. 4. Comparison of predicted and measured ammonia nitrogen levels at 3 sites across Poole Harbour and for 3 decay rates

CONCLUSIONS

16. The paper describes recent developments in the refinement of a numerical hydraulic and water quality model to predict tidal currents and nitrogen constituent levels in a large coastal wetland basin.

17. The finite difference model has been refined to include a more accurate representation of the complex hydrodynamic processes of flooding and drying of tidal wetlands. In these developments, not only were improvements made to the numerical treatment of these complex processes, but it was also established that Reynolds number effects could be significant in flow over wetlands, and hence the friction factor was related to both the relative roughness and the Reynolds number.

18. The model was also refined to include a higher order accurate representation of the advective process. Idealised tests showed that the ULTIMATE QUICKEST scheme was highly accurate and computationally efficient, and gave rise to no overshoot or undershoot (i.e. negative concentrations). In applying this scheme to Poole Harbour the scheme was first modified to include the flooding and drying processes, and was then found to give rise to close agreement with the field data at all sites across the basin. The refined model was then extended to consider the sensitivity of the results to varying and interactive rate constants, with this research still on-going at present.

ACKNOWLEDGEMENTS

19. The authors are grateful to the following organisations for funding and supporting this research project:- European Commission (contract no. CI1*0390-UK), UNIRAS A/S, NRA Wessex Region and Wessex Water plc.

REFERENCES

1. FALCONER R.A. Flow and water quality modelling in coastal and inland waters. Journal of Hydraulic Research, vol.30, no.4, 1992, 437-452.

2. CAHYONO. Three-dimensional numerical modelling of sediment transport processes in non-stratified estuarine and coastal waters. Thesis submitted in partial fulfilment for the degree of PhD, University of Bradford, Bradford, 1992, 315.

3. FALCONER R.A. A two-dimensional mathematical model study of the nitrate levels in an inland natural basin. Proceedings of the International Conference on Water Quality in the Inland Natural Environment, BHRA Fluid Engineering, Bournemouth, June 1986, 325-344.

4. FALCONER R.A. An introduction to nearly horizontal flows. The Coastal, Estuarial and Harbour Engineers Reference Book, (Eds. M B Abbott and W A Price), E & F N Spon, London, Chapter 2, 1993, 27-36.

5. HENDERSON F. M. Open channel flow. Collier-Macmillan Publishers, London, 1996, 522.

6. WU J. Wind stress and surface roughness at air-surface interface. Journal of Geophysical Research, vol.74, 1969, 444-455.

7. FALCONER R.A. and CHEN Y. An improved representation of flooding and drying and wind stress effects in a two-dimensional tidal numerical model. Proceedings of the Institution of Civil Engineers, Part 2, vol.91, December 1991, 659-678.

8. FISCHER H.B., LIST J.E., KOH R.C.Y., IMBERGER J. and BROOKS N.H. Mixing in inland and coastal waters. Academic Press Inc, San Diego, 1979, 483.

9. FALCONER R.A. Review of modelling flow and pollutant transport processes in hydraulic basins. Proceedings of First International Conference on Water Pollution: Modelling, Measuring and Prediction, Southampton, Computational Mechanics Publications, September 1991, 3-23.

10. BROWN L.C. and BARNWELL T.O. Computer program documentation for the enhanced stream water quality model QUAL2E. Environmental Research Laboratory, USA EPA, Athens GA, Report No. EPA/600/3-85-065, August 1985, 141.

11. FALCONER R.A. and OWENS P.H. Numerical simulation of flooding and drying in a depth averaged tidal flow model. Proceedings of the Institution of Civil Engineers, Part 2, vol.83, March 1987, 161-180.

12. STELLING G.S., WIERSMA A.K. and WILLEMSE J.B.T.M. Practical aspects of accurate tidal computations. Journal of Hydraulic Engineering, ASCE, vol.112, no.9, September 1986, 802-817.

Physical processes in tidal wetland restoration

P. GOODWIN, BSc, MS, PhD, CEng, MICE, MIWEM, PE, Technical Director, Philip Williams & Associates Ltd, San Francisco

SYNOPSIS. Extensive interest has been developed during the past decade in restoring, enhancing and protecting coastal wetlands to compensate for the dramatic global losses of these critical ecosystems.
 The hydroperiod, water quality and sedimentation characteristics, are the primary factors affecting the viability of a tidal wetland. Accurate numerical simulation of the hydrologic regime requires an understanding of the geomorphic variation of marsh channels in response to tidal flows and episodic flood events. Inlet channels connecting coastal lagoons/tidal wetland systems to the ocean are particularly susceptible to large variations in area and can be subject to periodic closure. Practical guidelines for accommodating each of these processes, developed for tidal wetlands on the West Coast of the United States, are described.

INTRODUCTION

1. Wetlands have been perceived generally as wastelands until the past few years. On a global basis, it has been estimated that over half of all wetlands have been lost (ref. 1). Wetlands have been destroyed by direct intervention which includes drainage for agriculture or landfill for development. Wetlands are also damaged by indirect consequences of development; including changes in the watershed hydrology, accelerated sedimentation and an increase in nutrient or pollutant loading. The US Fish and Wildlife Service has estimated that over 45,000 km^2 of wetlands in the US have been lost, or an area equivalent in size to Denmark (ref. 2). The loss of wetlands has not been uniform, for example during the past century Southern California has lost over 80% of all the historic coastal wetlands.

2. Wetlands influence regional biological diversity and cannot be considered as isolated ecosystems. For example, coastal wetlands may provide resting and feeding habitat for migratory bird flyways or provide critical gateways for salmon migration up rivers.

3. If a wetland is to be enhanced, restored or created, it is important to identify the significant physical processes occurring at the site. The changes to these physical processes as a result of the project or management plan must be predictable and compatible with the biological goals of the project. This paper

summarizes some of the significant physical processes that need to be considered in coastal wetland restoration and enhancement.

NUMERICAL SIMULATION

Governing Equations

4. The governing equations for tidal flows in coastal embayments are the equations for conservation of mass and momentum. Expressed in one dimensional form these equations are:

$$B\frac{\partial h}{\partial t} + \frac{\partial Q}{\partial x} = 0 \tag{1}$$

$$\frac{\partial Q}{\partial t} - \frac{2BQ}{A}\frac{\partial h}{\partial t} - B\frac{Q^2}{A^2}\frac{\partial h}{\partial x} = -gA\frac{\partial h}{\partial x} - \frac{g|Q|Q}{C^2 AR} \tag{2}$$

where t is time (s), x is the longitudinal distance coordinate (m), B is the channel width (m), Q is the discharge (m³/s), A is the cross-sectional area of flow (m²), h is the water surface elevation (m) and C is the Chezy resistance coefficient.

5. In tidal slough channel and wetland systems subject to significant cycles of deposition and erosion, it may be necessary to solve the equation for the conservation of sediment mass.

$$\frac{\partial}{\partial t}(cA) + \frac{\partial}{\partial x}(cQ) + (1-n)\,\rho_s B\frac{\partial z}{\partial t} = 0 \tag{3}$$

where c is the concentration of sediment in mass of sediment to volume of sediment-water mixture (kg/m³), z is the elevation of the bed (m), n is the bed porosity, and ρ_s is the density of the sediment (kg/m³).

6. One dimensional network models are appropriate for simulating flows in complex wetlands. For example, one of the earliest models was developed for Bolinas Lagoon (ref. 3).

7. In coastal lagoons with peripheral tidal marshes it is more appropriate to use the two dimensional depth integrated equations. The formulation has been given elsewhere (ref. 4),These equations can be solved by a range of standard numerical techniques (for example, ref. 4 and ref. 5).

8. The geomorphic features of San Elijo Lagoon are typical of many tidal wetland systems in California (Figure 1). A stream flows into a coastal lagoon, before discharging into the Pacific Ocean through a confined inlet channel. The inlet channel is subject to closure during periods of low flows in the streams or intense wave activity at the entrance to the inlet channel.

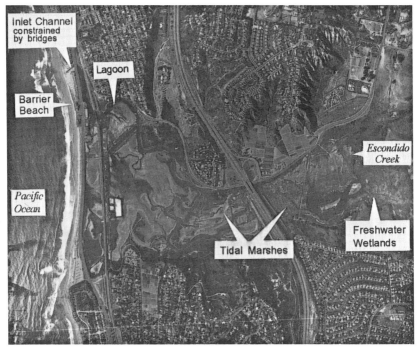

Figure 1. Geomorphic Features of San Elijo Lagoon, California

Closure Relations

9. When solving the governing equations, closure relations for C (or other resistance coefficient), c and turbulence closure relation for the two-dimensional formulation need to be determined. In addition, when applying these models to tidal wetlands there are additional difficulties which are more problematic than in open coast or large estuarine modeling applications, namely: wetting and drying of mudflats and marshplains; significant changes in bathymetry; fluctuating area of the inlet channel; and significant variations in the resistance across the marshplains.

Wetting and Drying

10. Many wetland restoration projects in California are undertaken to provide minimum areas of specific mitigation habitat. As a first approximation, salt marsh habitat appears to have a close correlation with the hydroperiod or representative water elevations. An example of the relation between reference tide elevation and habitat type is shown in Table 1. This data is based on information collected in San Francisco Bay and Los Angeles County.

Table 1. Elevation ranges for some typical California salt marsh habitats.

Habitat	Reference Tide Elevation
Upland	
	Extreme High Water (EHW)
Salt Grass	
	Mean Higher High Water (MHHW)
Pickleweed	
	Mean High Water (MHW)
Cordgrass	
	Mean Sea Level (MSL)
Mud Flat	
	Mean Lower Low Water (MLLW)
Sub Tidal	

11. Typical marshplains are flat (less than 1% slope). A 0.1 m error in predicted tide elevations could result in the projected habitat boundaries being over 10 m inaccurate. These boundaries are normally sinuous and follow the periphery of the channels on the marshplain. Therefore, inaccuracies in prediction of the hydrologic regime can result in large discrepancies in predicted habitat distribution. It is important to be able to simulate the wetting and drying of the marshplain accurately when obtaining estimates of the hydroperiod.

Size of Marsh Channels

12. The size of marsh channels adjust to the tidal prism of the upstream drainage, sedimentation rates, and tide range. If a marsh is created, restored or enhanced, it is useful to know how the tidal channels will adjust to the new conditions. Since the tidal prism may depend on the size of the marsh channels, there is a process of feedback developed. Sometimes it is possible to develop the channel sizes from a knowledge of the historic conditions of the site, or from a geomorphic model of an adjacent site. In the absence of local geomorphic information, empirical relationships are useful as a preliminary design tool.

13. Myrick and Leopold developed hydraulic geometry relationships for small tidal estuaries on the East Coast of the US (ref. 6). Haltiner and Williams (ref. 7) extended this work to obtain the marsh channel geometry (width, depth, cross-sectional area) as a function of the potential diurnal tidal prism. The potential diurnal tidal prism is defined as the tidal prism between MHHW and MLLW, assuming the water level in the marsh channel is identical to the tidal forcing. These empirical relationships, developed from a database of geomorphical characteristics of California wetlands may be expressed as:

$$B = 0.10 \, P_D^{0.49} \tag{4}$$
$$h = 0.16 \, P_D^{0.24} \tag{5}$$
$$A = 0.025 \, P_D^{0.61} \tag{6}$$

where B is the bank top width of the channel (m), h is the maximum depth of the channel (m), A is the cross-sectional area of the marsh channel (m²) and P_D is the potential diurnal tidal prism (m³).

14. These equations can be used to predict the size of the marsh channel that may be expected to develop following restoration or enhancement. These dimensions are only approximate and the marsh system will achieve its own condition of dynamic equilibrium. Some restoration projects have re-created marshplain elevations by placing dredged spoil material. However, it is important to ensure that the flows over the marshplain can generate shear stresses capable of eroding and forming the intricate system of marsh channels. Muzzi Marsh in San Francisco Bay is an example of a wetland restoration project with inhibited development of marsh channels, accompanied by sparse growth of salt marsh vegetation (ref. 20).

Size of Lagoon Inlet Channel

15. The equilibrium area of the inlet channel connecting a tidal lagoon or wetland to the ocean can be estimated by an equation of the form:

$$A_C = \alpha P^\beta \qquad (7)$$

where A_c is the cross-sectional area of the inlet channel measured below MSL (m²), and P is the tidal prism (m³). Some values for the coefficients α, β for natural inlet channels are summarized in Table 2 (refs. 8 - 15).

Table 2. Coefficients for Predicting Inlet Channel Area (A_c).

	α	β
O'Brien (1969) [2]	0.00081	0.85
O'Brien (1971) [2]	0.000066	1
Hume (1991) [1]	0.00044	0.915
Hume and Herdendorf (1993) [1]		
Barrier Enclosed	0.00025	0.927
River Mouth	0.0044	0.757
Johnson (1973) [3]	0.00006	1
LeConte (1905) [1]	0.00011	1
Nayak (1971) [2]	0.000062	1
Jarrett (1976) [2,4]	0.000038	1.03

[1] Potential prism based on spring tidal range
[2] Potential prism based on mean diurnal range
[3] Potential prism based on mean tide range
[4] Relationship based on no jetties or one jetty at the inlet

16. Equation 7 gives an estimate of the equilibrium cross-sectional area of the tidal inlet channel. The actual cross-sectional area will depend upon the balance between local cross-shore or longshore sediment transport and the scouring action of the tidal flows through the inlet. The cross-sectional area of the inlet in small tidal embayments can exhibit large changes in area from spring to neap tides, or even throughout the diurnal tidal cycle (Figure 2). These significant fluctuations in the inlet area can influence the hydrological response of the lagoon/wetland to tidal forcing (Section 20).

Flow through Vegetation

17. The resistance to flow in the marsh channels is controlled by the bed material roughness, bedform roughness and exchanges of flow between the

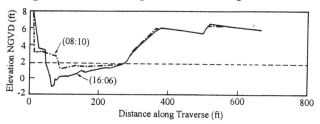

(a) Diurnal variations of inlet cross-section area, February 17, 1992

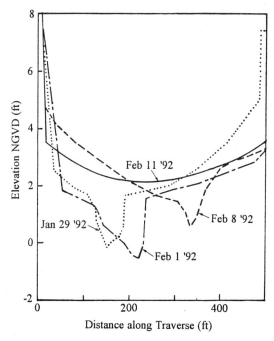

(b) Variation of inlet cross-section area at San Dieguito during a spring-neap cycle

Figure 2. Typical variations in a Southern California inlet channel. Source: H. Elwany (Scripps Institute of Oceanography, 1993)

marsh channel and marshplain. The resistance in the marsh channel and across mudflats can be characterized from synoptic tide and velocity measurements, and the resistance coefficients usually lie within a fairly narrow range. However, the resistance due to vegetation exhibits a greater range of values (Figure 3). The uncertainties in selecting an appropriate roughness coefficient is therefore also increased.

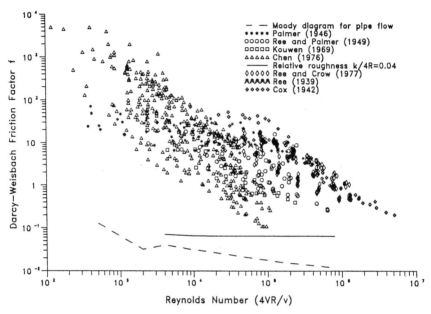

Figure 3. Resistance due to vegetation

18. Flows across marshplains may be classified into five regimes (ref. 16) as shown in Figure 4. The resistance will vary depending on the depth and velocity

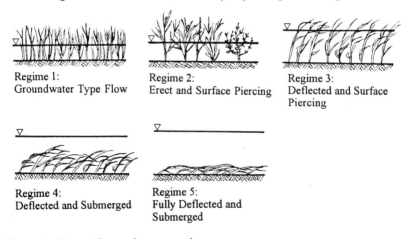

Figure 4. Vegetation resistance regimes

of flows across the marshplain. For example, pickleweed, a common marsh plant in California is characterized by a low resistance at small depths due to sparse stems and a complex network of micro channel beneath the plants. At greater depths, the flows encounter the dense succulent branches of the plant. Regime 5 is usually only experienced during periods of flood flows from creeks discharging into the marsh. Predictive equations for each regime are given in reference 16.

Closure Criteria

19. A preliminary estimate of whether the inlet channel of a lagoon or tidal wetland is subject to periodic closure may be obtained from considering the balance between longshore sediment transport and scouring ability of the inlet channel (ref. 12). Determination of the frequency of closure requires detailed modeling of the inlet channel dynamics.

Response to Tidal Forcing

20. Section 11 discussed the importance of an accurate prediction of the hydroperiod. It is therefore important to understand how the physical characteristics of the wetland are going to influence the tidal response of the system.

21. For a small lagoon/wetland system, the governing equations can be simplified (refs. 17, 18) by neglecting any freshwater inflow, assuming the water level in the lagoon and wetland is horizontal and the surface area of the water body is constant.

$$\frac{\partial Q}{\partial t} + \frac{\partial}{\partial x}\left(\frac{Q^2}{A_c}\right) = -gA_c \frac{\partial \eta}{\partial x} - \frac{fPQ|Q|}{A_c^2} \qquad (8)$$

$$Q = A_s \frac{d\eta}{dt} \qquad (9)$$

where f is the Darcy-Weisbach friction factor, η is the water surface elevation above datum (m), P is the wetted perimeter of the inlet channel (m), and A_s is the surface area of the lagoon/wetland (m^2).

22. Most tidal lagoon/wetland systems will exhibit damping of the tidal signal. The response of the simplified system (7) and (8) can be shown to result in a potential magnification of the tidal range in the lagoon for critical geometries and forcing frequencies (Figure 5).

HYDRAULIC PROCESSES

$$0.1 < \alpha = \frac{\omega}{\omega_n} = \omega\sqrt{\frac{LA_s}{gA_c}} > 20 \qquad (10)$$

$$\beta = \left(\frac{fLh}{A_c} + \text{entry and exit losses}\right)\frac{a_o A_s}{LA_c} < 10 \qquad (11)$$

where ω is the frequency of the tidal forcing and a_o is the amplitude of the ocean tide (m) and L is the length of the inlet channel (m).

23. From Figure 5, it may be seen that lagoons will respond to frequencies approaching the natural frequency of the system. However, high frequency forcing such as swells or wind waves at the inlet are subject to significant damping and the lagoon will not respond. This indicates that the depth-duration curve or hydroperiod is unlikely to be affected by wind generated waves. Exceptions to this may occur at the point of inlet closure, where wave overtopping of the bar formed in the inlet channel may raise the water surface level in the wetland.

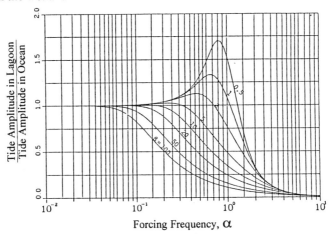

Figure 5. The fundamental response of a tidal lagoon to ocean forcing

24. The tidal wave form can be distorted by the generation of higher harmonics called shallow water tides or overtides (ref. 18). Overtides are due to local non-linear processes such as boundary resistance, bathymetric characteristics, and advection of momentum (ref. 19). The interaction of non-linear tidal components with the fundamental model results in residual velocities and mean water levels that are different from the ocean.

25. Flood tidal dominance occurs when the residual velocity is from the ocean toward the lagoon, and generally results in tidal set down in the lagoon and wetland. The mean water level in the lagoon is less than the ocean.

26. Ebb dominance occurs when the residual velocity is directed toward the ocean. This is associated with tidal set up in the wetland where the mean level in the lagoon exceeds the ocean.

27. Ebb or flood dominance is of importance when considering the maintenance requirements and geomorphic features likely to be experienced in the inlet channel. Ebb dominant channels resulted in a net flux of sediment out of the system and therefore will require less maintenance than a flood dominant system in wetlands with ephemeral freshwater inflow.

28. It may also be shown that vertical banks in a deep lagoon are more likely to result in flood dominance. Whereas small lagoons with extensive mudflats and intertidal zones tend to be ebb dominated with tidal set up.

29. The response of a typical coastal wetland to tides on February 17, 1992 is shown in Figure 6, using equations (7) and (8). Figure 6 shows that little damping occurs in small lagoons, provided a minimum cross-section area is

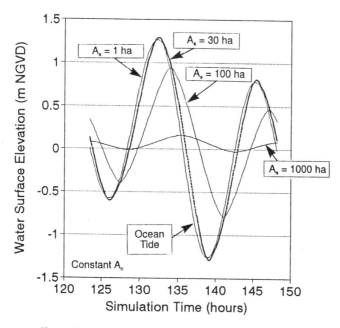

Figure 6. Effect of lagoon size on tidal response (February 17, 1992 tide)

maintained. Lagoons with a surface area of approximately 30ha exhibit some magnification of the tide, whereas in very large lagoons, significant damping is observed. The rates of sediment flux through the inlet for the sample conditions are shown in Figure 7. The flux of sediment is greater on ebb flows than flood flows. Figure 8 illustrates the results of a large tidal wetland comprising a subtidal lagoon of 140 ha, 4 ha of mudflats and 32 ha of marshplain.

HYDRAULIC PROCESSES

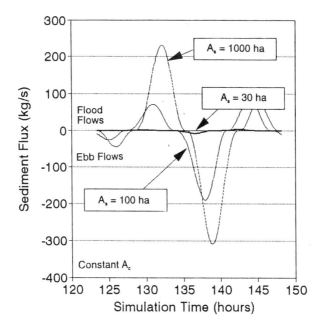

Figure 7. Sediment fluxes in tidal inlet for February 17, 1992 tide

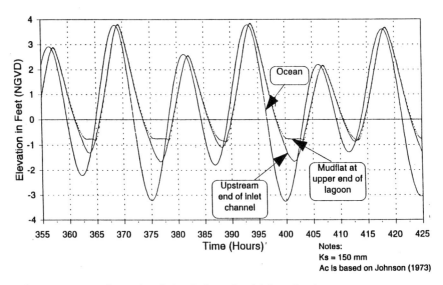

Figure 8. Two-dimensional simulation of a tidal wetland

30. A full two-dimensional model was used to generate these results (ref. 4), with a variable roughness coefficient and variable inlet geometry. Tidal setup is observed indicating an ebb dominant system. There is some amplification of the high tide elevations in the lagoon. These simulations show that it is possible to design a wetland configuration that maintains a large tidal range and will provide extensive intertidal habitat within the wetland.

Geomorphic Analyses

31. A tidal wetland represents a natural balance between sedimentation due to floods and tidal circulation. Design of a suitable tidal wetland project must be able to evolve in response to natural episodic events, and adjust to longterm sedimentation processes.

32. Significant episodic events include periodic flooding, delivering large sediment loads and altering river and marsh channels, large wave storm events, barrier beach migrating inland and reduction in tidal prism due to blocking of major marsh channel. These events have been documented in the Tijuana River Estuary (ref. 20). Conversely the 1993 storms increased the tidal prism of San Dieguito Lagoon from $0.225 \times 10^6 m^3$ to $0.340 \times 10^6 m^3$.

CONCLUSIONS

33. Accurate characteristics of the hydrologic regime of a tidal wetland requires a detailed understanding of the marsh channel geometry, the inlet channel response to wave and tide conditions, the flow resistance on the marshplain, and the propagation of the wetting or drying front.

34. The sensitivity of each of these factors to the hydrologic characteristics of tidal wetlands should be assessed in the restoration or enhancement of tidal wetlands.

REFERENCES

1. MALTBY, E. Waterlogged wealth: why waste the worlds wet places? Earthscan, 1986, p.200.
2. TINER, R.W. Wetlands of the United States: Current status and trends. US Fish and Wildlife Service, 1984.
3. FISCHER H.B. A Lagrangian method for predicting pollutant transport in Bolinas Lagoon, California. US Geological Survey Open File Report, Menlo Park, California, 1969.
4. FALCONER R.A. A two-dimensional mathematical model study of the nitrate levels in an inland natural basin. International Conference on Water Quality Modeling in the Inland Natural Environment, BHRA, Bournemouth, 1986, 325-344.
5. ABBOTT M.B. and CUNGE J.A. (eds.). Engineering applications of computational hydraulics: Homage to Alexandre Preissmann. Pitman, p.224.
6. MYRICK R.M. and LEOPOLD L.B. Hydraulic geometry of a small tidal estuary. US Geological Survey, 1963, Professional Paper 422-B, p.18.

7. HALTINER J.P. and WILLIAMS P.B. Hydraulic design in salt marsh restoration. Proceedings of the National Symposium on Wetland Hydrology, Association of State Wetland Managers, Chicago, Illinois, 1987.
8. O'BRIEN M.P. Equilibrium areas of inlets on sandy coasts. Journal of Waterways, Harbors and Coastal Engineering Division, ASCE, 1969, vol. 95, no. WW1, February, 43-51.
9. O'BRIEN M.P. Notes on tidal inlets on sandy shores. Hydraulic Engineering Laboratory, University of California, Berkeley, 1971, Report HEL-24-5, p.52.
10. HUME T.M. Empirical stability relationships for estuarine waterways and equations for stable channel design. Journal of Coastal Research, 1991, vol. 7 no. 4, 1097-1111.
11. HUME T.M. and HERDENDORF C.E. On the use of empirical stability relationships for characterizing estuaries. Journal of Coastal Research, 1993, vol. 9, no. 2, 413-420.
12. JOHNSON J.W. Characteristics and behavior of Pacific Coast tidal inlets. Journal of Waterways, Harbors and Coastal Engineering Division, ASCE, 1973, vol. 99, no. WW3, August, 325-339.
13. LeCONTE L.J. Discussion of "Notes on the Improvement of River and Harbor Outlets in the United States" by D.A. Watt. Transcripts of the ASCE, 1905, vol. LV, December, 306-308.
14. NAYAK I.V. Tidal prism-area relationships in a model inlet. Hydraulic Engineering Laboratory, University of California, Berkeley, 1971, Report HEL 24-1.
15. JARRETT J.T. Tidal prism-inlet area relationships. GITI Report 3, US Army Corps of Engineers, Coastal Engineering Research Center, Fort Belvoir, Virginia, and Waterways Experiment Station, Vicksburg, Mississippi, 1976.
16. LEWANDOWSKI J.A. Vegetation resistance and circulation modeling in a tidal wetland. Dissertation submitted in partial satisfaction of the requirements for the degree of Doctor of Engineering, Hydraulic Engineering Laboratory, University of California, Berkeley, 1993.
17. KEULEGAN G.H. Tidal flows in entrances: Water level fluctuations of basins in communications with seas. US Army Corps of Engineers Waterways Experiment Station, Vicksburg, Mississippi, Technical Bulletin No. 14, 1967.
18. DiLORENZO J.L. The overtide and filtering response of small inlet/bay systems. In: Hydrodynamics and Sediment Dynamics of Tidal Inlets. D.G. Aubrey and L. Weishar (eds), Springer-Verlag, 1988, 24-53.
19. SPEER P.E. and AUBREY D.G. A study of non-linear tidal propagation in shallow inlet/estuary systems, Part II: Theory. Est. Coastal Shelf Sci., 1985, vol. 21, 207-224.
20. GOODWIN P. and WILLIAMS P.B. Restoring coastal wetlands: The California experience. Journ. of IWEM, 1992, vol. 6, no. 6, 709-719.

An integrated eco and hydrodynamic model for prediction of wetland regime in the Danubian Lowland

J. C. REFSGAARD, Chief Hydrologist, and K. HAVNO, Head of River Hydraulics Division, Danish Hydraulic Institute, and J. K. JENSEN, Senior Biologist, Danish Water Quality Institute

SYNOPSIS. An integrated mathematical modelling system describing flows, water quality processes, sediment transport/erosion in river, flood plain, reservoir and ground water system is being developed and fully coupled with a large data base and GIS system. The modelling and information system is being established in an ongoing project "Danubian Lowland - Ground Water Model". The system will be applied for assessing environmental impacts on eg ground water and floodplains of alternative water management strategies in the area, including alternative operation plans for the Gabcikovo hydropower scheme. The present paper gives a brief description of the established modelling and information system with special emphasis on the coupling between MIKE SHE and MIKE 11 and of the plans for model application during the project with special emphasis on aspects relating to wetland hydrology.

DANUBIAN LOWLAND - BACKGROUND
1. The Danubian Lowland between Bratislava and Komárno is an inland delta formed in the past by river sediments from the Danube. The entire area forms an alluvial aquifer, which throughout the year receives in the order of 25 m^3/s infiltration water from the Danube in the upper parts of the area and returns it into the Danube and the drainage channels in the downstream part. The aquifer is an important water resource for municipal and agricultural water supply.
2. Human influence has gradually changed the hydrological regime in the area. Construction of dams upstream of Bratislava together with exploitation of river sediments has significantly deepened the river bed and lowered the water level in the river. These changes have had a significant influence on the conditions of the ground water regime as well as the sensitive riverside forests downstream of Bratislava. In spite of this basically negative trend the floodplain area with its alluvial forests and the associated ecosystems still represents a very unique landscape of outstanding importance.
3. The construction of the hydraulic structures in connection with the hydropower plant at Gabcikovo also significantly affects the hydrological regime and the ecosystem of the region.
4. Industrial waste and municipal sewage from Bratislava and its surroundings together with the diffuse sources of agricultural fertilizers and agrochemicals are polluting the rivers, soil and ground water.
5. These physical and biochemical changes may reduce the atmospheric oxygen transport to the ground water and at the

same time increase the supply of organic matter which will change the oxidizing conditions to reducing conditions and thereby seriously deteriorate the ground water quality.

6. Due to the economical and ecological importance of the area comprehensive data collection programmes have been carried out for many years and a large number of studies have been made in the past. Some of the present environmental problems are published in (ref 1-2).

7. To utilize state-of-the-art modelling technology for addressing the water resources problems in the area the project "Danubian Lowland - Ground Water Model" has been defined within the PHARE programme agreed upon between the Commission of the European Communities and the Government of the Slovak Republic.

OBJECTIVE AND FRAMEWORK OF THE PHARE PROJECT

8. To understand and analyze the complex relationships between physical, chemical and biological changes in the surface- and subsurface water regimes requires multidisciplinary expertise in combination with advanced mathematical modelling techniques.

9. The overall project objective is to establish a reliable impact assessment model for the Danubian Lowland area, which enables the authorities to formulate optimal management strategies leading to a protection of the water resource and a sound ecological development for the area.

10. The PHARE project is being executed by the Slovak Ministry of the Environment. Specialists from the following Slovakian organisations are involved in various aspects of the project implementation:

* Comenius University, Faculty of Natural Science (PRIF UK)
* Water Research Institute (VUVH)
* Irrigation Research Institute (VUZH)

A Danish-Dutch consortium of six organisations was selected as consultant for the project.

11. The project was initiated in the beginning of 1992 and will be completed by the end of 1995. At present the necessary equipment has been delivered to the project in Bratislava and the modelling and information system has been installed on computer workstations. The modelling work has been initiated; however no final model calibrations nor predictions have been carried out at this stage.

ESTABLISHMENT OF DANUBIAN LOWLAND INFORMATION SYSTEM (DLIS)

12. An automated system is presently being developed to support the modelling activities. The integrated modelling system will be interfaced to a central information system. The central information system, called Danubian Lowland Information System (DLIS), will provide the different models with the necessary data and functionality to elaborate further on the modelling results.

13. Because of the complexity and the amount of data involved - first estimates indicate about 2 Gbyte of data - major attention has been paid to the development of the underlying data model. The information needed for the modelling originates from different monitoring networks, which are maintained and observed by different institutes. The larger part of the data is available from automated archives and can be loaded into the system from magnetic medium.

14. The information to be incorporated in DLIS has a pronounced spatial character. The two main components of the GIS are a geographical information system (GIS), for which ARC/INFO has been selected, and a relational data base management system (RDBMS), for which INFORMIX has been selected.

ESTABLISHMENT OF INTEGRATED MODELLING SYSTEM

15. The integrated modelling tool, which will form the basis for all the modelling activities, is based on the following packages which can be used individually or brought together in an integrated manner:

* **MIKE SHE** which, on catchment scale, can simulate the major flow and transport processes of the hydrological cycle which are traditionally divided in separate components:
  - 1-D flow and transport in the unsaturated zone
  - 3-D flow and transport in the ground water zone
  - 2-D flow and transport on the ground surface
  - 1-D flow and transport in the river.

  All the above processes are fully coupled allowing for feedbacks and interactions between components. In addition to the above mentioned components, MIKE SHE includes modules for multi-component chemical reactions in the unsaturated and saturated zone as well as a component for oxygen consumption and transport in the unsaturated zone.

* **MIKE 11**, which is a one-dimensional river modelling system. MIKE 11 is used for hydraulics, sediment transport and morphology, and water quality. The modules for sediment transport and morphology are able to deal with cohesive and non-cohesive sediment transport, as well as the accompanying morphological changes of the river bed. The non-cohesive model will operate on a number of different grain sizes, taking into account shielding effects. The cohesive model deals both with consolidation of the river bed and flocculation.

* **MIKE 21**, which is a two-dimensional hydrodynamic modelling system. MIKE 21 is used for reservoir modelling, including hydrodynamics, sediment transport and water quality. The sediment transport modules deals with both cohesive and non-cohesive sediment, and the non-cohesive module will operate on a number of different grain size fractions.

* Both of the above mentioned models include **River/Reservoir Water Quality (WQ) and Eutrophication (EU)** modules to describe oxygen, ammonium, nitrate and phosphorus concentrations and oxygen demands as well as eutrophication issues.

* **DAISY** is a one-dimensional root zone model for simulation of crop production, soil water dynamics, and nitrogen dynamics in crop production for various agricultural management practices and strategies. The particular processes considered include transformation and transport involving water, heat, carbon and nitrogen.

16. The above mentioned models are all generalized tools with comprehensive applicability ranges, and they are well proven in a large number of international projects. In addition, some model modifications are being carried out during the project in order to accommodate the very special

HYDRAULIC PROCESSES

environment and problems observed in the area.

17. With regard to simulation of floodplain hydrology and ecology the core of the integrated modelling system is constituted by the MIKE SHE, the MIKE 11 and a newly developed, full coupling of the two systems as described in the following three sections.

MIKE SHE

18. The European Hydrological System - SHE was developed in a joint effort by Institute of Hydrology (UK), SOGREAH (France) and Danish Hydraulic Institute. It is a deterministic, fully-distributed and physically-based modelling system for describing the major flow processes of the entire land phase of the hydrological cycle. A description of the SHE is given in (Ref. 3-4). Since 1987 the SHE has been further developed independently by the three respective organizations which now are University of Newcastle (UK), Laboratoire d'Hydraulique de France and DHI. DHI's version of the SHE, known as the MIKE SHE, represents significant new developments with respect to user interface, computational efficiency and process descriptions.

19. MIKE SHE solves the partial differential equations for the processes of overland and channel flow, unsaturated and saturated subsurface flow. The model is completed by a description of the processes of snow melt, interception and evapotranspiration. The flow equations are solved numerically using finite difference methods.

20. In the horizontal plane the catchment is discretized in a network of grid squares. The river system is assumed to run along the boundaries of these. Within each square the soil profile is described in a number of nodes, which above the groundwater table may become partly saturated. Lateral subsurface flow is only considered in the saturated part of the profile. Fig. 1 illustrates the structure of the MIKE SHE. A description of the methodology and some experiences of model application are presented in (Ref. 5-8).

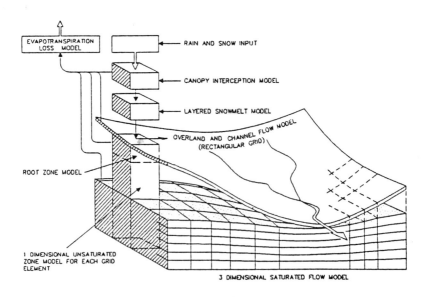

Fig. 1. Schematic presentation of the MIKE SHE.

21. The spatial and temporal variations in the catchment characteristics and meteorological input are provided in a series of two-dimensional matrices of grid square codes. To each code is further attached a number of attributes describing either parametric data or input data.

22. The distributed description in the MIKE SHE allows the user to include and test against spatially varying data. MIKE SHE is a multi output model which besides discharge in any river link also produces information about e.g. water table elevations, soil moisture contents, infiltration rates, evapotranspiration, etc. in each grid square.

MIKE 11

23. MIKE 11 is a comprehensive, one-dimensional modelling system for the simulation of flows, sediment transport and water quality in estuaries, rivers, irrigation systems and other water bodies. It is a 4th generation modelling package designed for micro-computers with DOS or UNIX operating systems and provides the user with an efficient interactive menu and graphical support system with logical and systematic layouts and sequencing in the menus. The package was introduced in 1989 and today the number of installations world-wide exceeds 300. The modular structure of MIKE 11 is illustrated in Fig. 2.

24. The hydrodynamic module of MIKE 11 is based on the complete partial differential equations of open channel flow (Saint Venant). The equations are solved by implicit, finite difference techniques. The formulations can be applied to branched and looped networks and quasi two-dimensional flow simulations on floodplains.

25. MIKE 11 operates on the basis of information about the river and the floodplain topography, including man-made hydraulic structures such as embankments, weirs, gates, dredging schemes and flood retention basins. The hydrodynamic module forms the basis for morphological and water quality studies by means of add-on modules.

A COUPLING OF MIKE SHE AND MIKE 11

26. The focus in MIKE SHE lies on catchment processes with a comparatively less advanced description of river processes. In contrary MIKE 11 has a more advanced description of river processes and a simpler catchment description than MIKE SHE. Hence, for cases where full emphasis is needed for both river and catchment processes a coupling of the two modelling systems is required.

27. A full coupling between MIKE SHE and MIKE 11 has been developed. MIKE 11 computes water levels and flows in the river and floodplain system. The water levels, flows and flooded areas from MIKE 11 are then used as boundary conditions in MIKE SHE for calculation of the remaining parts of the hydrological cycle. The interactions between the river and the other components (aquifer, overland flow, etc) computed in MIKE SHE are then in turn transmitted back to MIKE 11.

28. The two systems are run simultaneously with full exchange of data. Numerically, the two systems may utilize different time steps. The data transfer between the two systems takes place through shared memory.

29. The MIKE SHE-MIKE 11 coupling is crucial for a correct description of the dynamics of the river-aquifer interaction. Firstly, the river width is larger than one MIKE SHE grid, in which case the MIKE SHE river-aquifer description is no longer valid. Secondly, the river/reservoir system comprises

HYDRAULIC PROCESSES

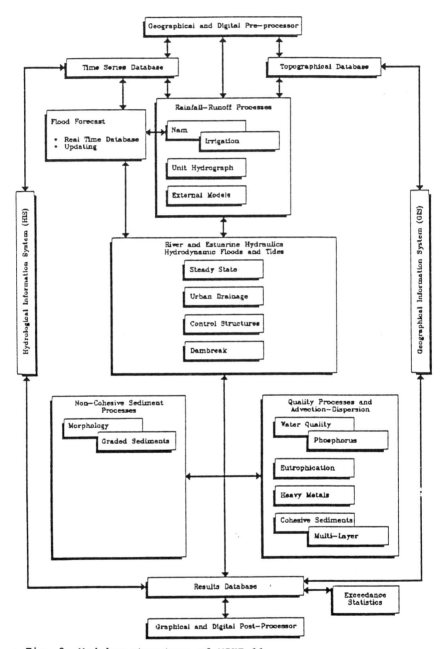

Fig. 2. Modular structure of MIKE 11

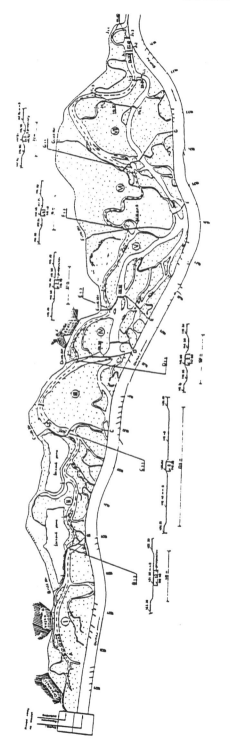

Fig. 3. Sketch of river branch system on the Slovakian floodplain for a reach of 20 km downstream of the reservoir

HYDRAULIC PROCESSES

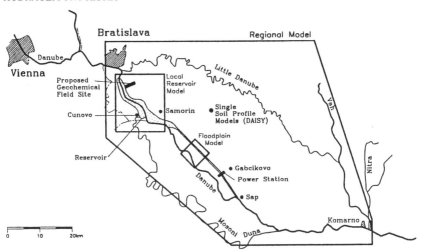

Fig. 4. Index map of the Danubian Lowland with indications of various spatial modelling scales.

a large number of hydraulic structures, the operation of which cannot be accounted for in MIKE SHE. Thirdly, the very complex branch system with loops and flood cells needs a very efficient hydrodynamic formulation such as MIKE 11's.

30. The complexity of the floodplain with its river branch system is shown in Fig. 3 for the 20 km reach downstream the reservoir on the Slovakian side where alluvial forest occurs. In order to enable predictions of possible changes in floodplain ecology it is crucial to provide a detailed description of both the surface water and the groundwater systems in this area as well as of their interaction. For this purpose the MIKE SHE-MIKE 11 coupling is required.

MODELLING STUDIES IN THE DANUBIAN LOWLAND

31. The modelling studies initiated and planned under the PHARE project involve a number of disciplines and processes with different space and time scales as outlined below. An index map also illustrating the different spatial scales is shown in Fig. 4.

32. River and Reservoir Flow and Sediment Transport. For the simulations of flows and sediment transport in the reservoir and the old Danube a combination of one and two-dimensional numerical models is applied.

33. MIKE 11 is being established for the Danube from a point within Austria to Komárno. It includes possibilities for imposing specific operation of the structures in connection with the hydropower plant at Gabcikovo and the reservoir weir at Cunovo. For the old Danube reach between Cunovo and Sap the complex channel branch system with all its internal regulating structures is included together with possibilities for directly describing inundation depths and coverage of the flood plains in between the branches. Thus the model is able to describe both low and high flow conditions as well as all possible regulation possibilities.

34. For the regional ground water studies the Little Danube and the irrigation and drainage systems are also included in the model setup.

35. The model will be calibrated on conditions from the 1960's as well as on the present conditions.

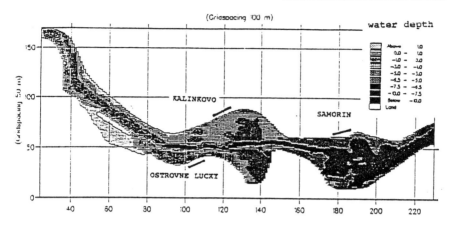

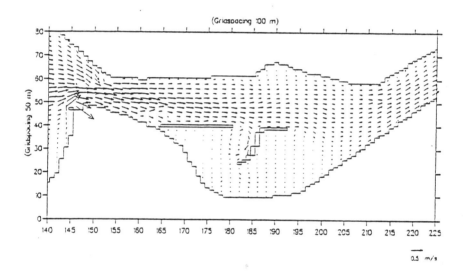

Fig. 5. Example of MIKE 21 model prediction of flow pattern in the Samorin reservoir with the designed reservoir alignment and deflecting structures. On the upper figure the bathymetry for the entire reservoir is shown together with the location of three major well fields for water supply. The lower figure shows the simulated velocities corresponding to a discharge of 1400 m3/s into the reservoir and power canal.

36. Long term morphological simulations will be carried out in order to assess bed level changes and composition of sediment in the backwater zone of the reservoir and in the old Danube (due to e.g. flushing of sediment from the reservoir and possible dumping of dredged material).
37. Hydrodynamic modelling with MIKE 21 has already been carried out for different options of reservoir alignments and deflecting structure designs, see Fig 5.

HYDRAULIC PROCESSES

38. The sediment transport modelling of suspended load will include both cohesive and non-cohesive transport. Also resuspension during flood of sediment deposited in periods with low flow will be accounted for. The boundary conditions in terms of flux and sediment transport at the upstream boundary, water level/flux at the entrance to the power channel, water level/flux at the weir will be provided from the MIKE 11 simulations. Sediment transport boundary conditions will consist of grain size distributions and suspended sediment concentration for each fraction.

39. The predictions will provide information about flows, water levels, grain size distributions and depths of the deposited suspended sediments in the reservoir. These results will be used both in the surface water quality modelling as well as in the ground water quality and quantity modelling. For assessment of the morphological consequences of different flushing schemes, the MIKE 21 non-cohesive sediment transport module will be applied.

40. <u>Surface Water Quality</u>. In order to highlight the oxygen status of the surface water, particularly during low flow situations and in slow flowing branches of the Danube, MIKE 11WQ is applied. The MIKE 11WQ describes the diurnal variation in the water quality parameters, i.e. the concentration of organic matter, oxygen, ammonium and nitrate in the water. The diurnal variation is especially important in the branches of the old Danube with a significant macrovegetation, in areas where severe eutrophication occurs and in periods with relatively low flow velocities.

41. The model will be used both in the old Danube with its branches as well as in the reservoir (here as a submodel to MIKE 21). Output from the eutrophication model, (which describes the daily average production), with respect to levels of oxygen production from primary producers, can be used in these calculations.

42. A description of the horizontal differences in the algae growth and the possible sedimentation of organic matter in the reservoir will be carried out by using the two-dimensional MIKE 21EU. Because of the relatively shallow areas in the reservoir macrophytes should be included in the future eutrophication modelling.

43. The eutrophication effect in the old Danube and the branch system will be most severe in periods with low flow. In these periods the system can be described by a branched 1-D system because the old Danube will be described by a branched/looped MIKE 11 model. The eutrophication effects can also be described by the one dimensional MIKE 11EU-model including macrophytes.

44. <u>Ground water flow modelling</u>. The application of MIKE SHE at three spatial scales, see Fig. 3, will support different types of modelling studies:

* **Regional scale.** The aim at this scale is to provide a framework for regional predictions and provide realistic boundary conditions (usually head boundaries) for local models. Some management options could have subregional or even regional implications requiring a reliable model on this scale. For the purposes of detailed modelling in local areas, where a very detailed description is required, it is only necessary and only technically feasible to establish a model for a smaller area. If the exact boundary conditions are not easily established, the regional model can provide the dynamic boundary conditions

which may account for the conditions outside the model area.

The regional model includes the entire Zitny Ostrov area and cover approximately 3000 km². The overall hydrological regime will be simulated taking into account all the major surface water systems.

* **Local scale.** For studies of the local conditions MIKE SHE will be set up in small areas to describe the flow and transport conditions with a high degree of detail both in horizontal and vertical directions (a fully 3-D description). These models will serve as a basis for the detailed description of different aspects, e.g.:
  - Geochemical processes around the reservoir area; and
  - ecological effects in the Danubian flood plain area.
* **Transect/Plot scale.** Model simulations on this scale will basically serve to study specific processes which either require a very fine spatial resolution or which can be described by one-dimensional flows, e.g.:
  - Geochemical processes (e.g. along a transect in connection with field investigations);
  - flow and solute (e.g. nitrate) processes in the unsaturated zone (soil columns); and
  - analysis of oxygen transport from the atmosphere through the unsaturated zone to the water table (transect or soil column).

45. Geochemical modelling. The hydrogeochemical modelling focuses on the part of the aquifer system in the vicinity (2-3 km) of the Danube river.

46. The infiltration of oxygen and nitrate through river (and reservoir) beds with different compositions, e.g. fine sediments rich in organic material and gravel sediments with a small amount of organic material, will result in different supplies of oxidation capacity to the aquifer system at the different river-aquifer interfaces. In some parts of the interface between river and aquifer, water with a low oxidation capacity and perhaps even anoxic will infiltrate due to the build up of a low permeable layer of fine sediment in the river after the completion of the reservoir. It is therefore important to handle the transfer of oxidation capacity through various river bed systems correctly in order to be able to characterize the supply of oxidation capacity to the aquifer system. Oxidation capacity can also be added to the system through infiltration from the unsaturated zone. Further, the effects of a fluctuating water table in the riverine area in bringing oxygen (and nitrate) to the aquifer will be studied.   47.   The hydrogeochemical modelling will focus on the oxidation/reduction processes in the river bed and aquifer systems. The total amount of oxygen and nitrate is equal to the oxidation capacity added to the systems through the river and unsaturated zone (SO_4-reduction has not been considered here). Bulk organic matter, either in dissolved or solid form, is the reduction capacity. The hydrogeochemical situation will be that of a reduction of oxygen and nitrate by kinetically-controlled oxidation of organic matter. Oxygen must be consumed first before nitrate is reduced.

48. Unsaturated zone and agrochemical modelling. The effect of the reservoir on the productivity in the Zitny Ostrov is of direct concern. From the calibrated regional

HYDRAULIC PROCESSES

model the present and future ground water levels will be simulated and the area can be classified according to its water supply for crops in the growing season.

49. Combining these predictions with the DAISY agricultural model the productivity and the irrigation needs before and after implementation of a given management scenario for the dam can be simulated for selected classes of water table depths and crops.

50. Although the amounts of nitrate leached are not extremely large, they are a problem for the general quality of the ground water. The losses can be reduced through changes in amounts of N applied, timing of application, optimal use of manure, and by optimal irrigation practices. Different scenarios can be analyzed through modelling.

51. Simulation with DAISY (Ref.9) can provide estimates of former and future nitrate loads leaching from the root zone under different conditions and to map "leaching hazard". Through discussions with the relevant agricultural institutions a number of scenarios can be defined with improved agricultural systems/protected areas and they can analyze how this would influence the leaching losses and ground water quality.

52. <u>Modelling Ecological Effects in the Flood Plain</u>. The ecological functioning of the floodplain is governed by the dynamics of inundation, flushing and ground water level fluctuations. These factors will form part of the modelling of the floodplain area.

53. The MIKE SHE/MIKE 11 model will be set up for an area which forms part of the existing monitoring system (for biomonitoring and forestry).

54. The model will be given sufficient detail in order to simulate the inundation and flushing regime at various discharges in the old Danube in order to predict changes in ecotype diversity. The horizontal model discretization is envisaged to be in the order of 50 m. The model will also allow the prediction of ground water levels, soil moisture regime of the floodplain in relation to channel and river branch development (e.g. morphology and sedimentation).

55. Water quality aspects will be considered as being included using MIKE 11 WQ/EU, thus enabling predictions of the water quality and eventually macrophyte growth.

CONCLUSIONS

56. The ecological system of the Danubian Lowland is so complex with so many interactions between the surface and the subsurface water regimes and between physical, chemical and biological changes that a comprehensive mathematical modelling system is required in order to provide quantitative assessments of environmental impacts.

57. Such modelling system coupled with a comprehensive data base/GIS system is being developed under the PHARE project. When finally calibrated and verified this modelling and information system will provide the best available tool for providing assessments of the impacts on surface and ground water quantity and quality of alternative water management schemes.

58. In addition, the integrated system will enable detailed, quantitive predictions of surface and ground water regime in the floodplain area, including e.g. frequency, magnitude and duration of inundations. Such information constitutes a necessary basis for subsequent analysis of flora and fauna in the floodplain.

ACKNOWLEDGEMENT
59. The PHARE project is being executed by the Slovak Ministry of the Environment. The Project Manager is Professor Igor Mucha from the Ground Water Consultants, Faculty of Natural Science, Comenius University (PRIF UK). A Danish-Dutch consortium of six organizations was selected as Consultant for the project. The Consultant is headed by Danish Hydraulic Institute (DHI) and comprises the following associated partners: DHV Consultants BV, The Netherlands; TNO-Applied Institute of Geoscience, The Netherlands; Water Quality Institute (VKI), Denmark; I Krüger Consult AS, Denmark; and the Royal Veterinary and Agricultural University, Denmark. The author of the present paper is the Team Leader for the Consultant.

REFERENCES
1. MUCHA I. and PAULIKOVA E. "Ground water quality in the Danubian lowland downwards from Bratislava". European Water Pollution Control, Vol 1, 5, 13-16, 1991.
2. SOMLYODY L. and HOCK B. " State of the water environment in Hungary". European Water Pollution Control, Vol 1, 1, 43-52, 1991.
3. ABBOTT M.B., BATHURST J.C., CUNGE J.A., O'CONNELL P.E. and RASMUSSEN J. "An Introduction to the European Hydrological System - Système Hydrologique Européen "SHE" 1: History and philosophy of a physically based distributed modelling system". Journal of Hydrology, 87, 45-59, 1986.
4. ABBOTT M.B., BATHURST J.C., CUNGE J.A., O'CONNELL P.E. and RASMUSSEN J. "An Introduction to the European Hydrological System - Système Hydrologique Européen "SHE" 2: "Structure of a physically based distributed modelling system". Journal of Hydrology, 87, 61-77, 1986.
5. REFSGAARD J.C., SETH S.M., BATHURST J.C., ERLICH M., STORM B., JØRGENSEN G.H. and CHANDRA S. "Application of the SHE to catchment in India - Part 1: General results". Journal of Hydrology, 140, 1-23, 1992.
6. JAIN S.K., STORM B., BATHURST J.C., REFSGAARD J.C. and SINGH R.D. "Application of the SHE to catchments in India - Part 2: Field experiments and simulation studies on the Kolar Subcatchment of the Narmada River". Journal of Hydrology, 140, 25-47, 1992.
7. LOHANI V.K., REFSGAARD J.C., CLAUSEN T., ERLICH M. and STORM B. "Application of the SHE for irrigation command area studies in India". Journal of Irrigation and Drainage Engineering, 119 (1), 34-49, 1993.
8. STYCZEN M. and STORM B. "Modelling of N-movements on catchment scale - a tool for analysis and decision making. 1. Model description & 2. A case study". Fertilizer Research, 36, 1-6, 7-17, 1993.
9. HANSEN S., JENSEN H.E., NIELSEN N.E. and SVENDSEN H. "Simulation of nitrogen dynamics and biomass production in winter wheat using the Danish simulation model DAISY". Fertilizer Research, 27, 245-259, 1991.

The role of computational models in wetland management – a case study from Bangladesh

Dr C. E. REEVE, Head, Integrated Environmental Management, and
Dr N. WALMSLEY, Project Manager, Overseas Development Unit,
HR Wallingford

SYNOPSIS. Surface water flow models are specifically orientated to provide water level and discharge time series at nodes within the modelled river system. This information is used extensively to assist in the engineering design of flood control measures. These models may also be used to assess the impacts away from the main river courses on the floodplains. A suite of post-processing programs were developed to provide additional data and information relating to floodplains. In particular these were to assess the impact flood control measures would have on :

- the timing, duration and depth of floodplain inundation
- the change this would introduce on areas suitable for fish.

This paper describes the methodology of the approach taken and illustrates the application of the techniques by examples from the North West and South East Regional Studies in Bangladesh. The techniques developed are a useful tool to provide additional primary and secondary data for experts in disciplines such as wetland management, agriculture, fisheries, environmental impact assessment, sociology and economics. The analysis is also of considerable benefit to modellers since it serves as an additional verification of the model's ability to predict flooding regimes in areas away from the river system, that is, on the active floodplains.

INTRODUCTION
1. Surface water flow models are specifically orientated to provide water level and discharge time series at nodes within the modelled river system. This information is used extensively to assist in the design of river engineering and flood control measures.
2. In river basins which experience regular inundation over their floodplains floods impose an important, often dominant, constraint on the ecology of the floodplains. Flood control measures in these areas relie on achieving a fine balance between engineering based solutions and the impacts these may have on the development of the floodplains together with the dis-benefits which may be introduced through loss of fisheries potential

and other possible negative impacts on the environment.

3. Typical of these types of flooded areas is the Bangladeshi delta. Up to 60% of the country is inundated annually by riverine flooding.

4. Post-processing of the surface water flow model results can lead to additional data and information related to the floodplains. This can be used to assist in the evaluation and interpretation of non-engineering aspects such as, ecology, agricultural potential, fisheries, environmental impacts, and socio-economic evaluations.

5. The post-processing relies on using the model results together with topographic data relating to the ground elevation. These enable flood depths and durations on the floodplain to be calculated.

METHODOLOGY

Long term simulations

6. Many flood studies undertaken using computational models use design events (input discharges and tailwater levels) of a specified return period for design purposes. However, in very complex hydraulic regimes where rainfall, upland discharges and tailwater level all play an important role the specification of design events may not be possible.

7. Where data is available, long time series simulations can provide sufficient information from which to estimate the return period levels and discharges at specified points within the model system, thereby avoiding the need for complex joint probability calculations.

8. Due to the complexity of the Bangladeshi Delta and the interaction of the various flood causing factors, the definition of design events of a given return period in terms of standardised boundary conditions is impossible. In an attempt to overcome these problems a rationale was used which involved long term simulations of hydraulic models over a period of 25 years. In detail the rationale required,

- the preparation of boundary conditions required to run models for the period 1965-89.
- running the models for the full 25 year period, at least once for the present (baseline) conditions and once for the ultimately adopted scheme(s).
- combinations of various options to reach the final plan may be studied on the basis of simulations for a reduced number of selected seasons, the selections being based on the analysis of the 25 year baseline run.
- sensitivity analysis of ultimately adopted scheme considering changed boundary conditions in the major rivers due to proposed schemes outside the region.
- statistical analysis of the results, aimed at assigning return periods to historic peak, seasonal or sub-seasonal values of selected design variables.

Topographic data

9. The primary objective of the post-processing is to predict the depth and duration of flooding over the floodplain. This requires water level predictions from the computational model and also details of the ground elevations.

10. The floodplain areas are sub-divided into a series of cells in which the flood level is represented by a water level node within the computational model set-up. In each cell, the area/elevation characteristics of the ground elevation are obtained from spot heights from suitable topographic maps. The finer the resolution of the spot heights the greater the accuracy of the area/elevation curve. The water levels together with the area/elevation curves for the floodplains allows the flood depths and durations to be calculated for the floodplain area.

Flood / depth phases

11. In Bangladesh, the land is often categorised by reference to its flood phase. The flood phase is a measure of the maximum depth of flooding in a flood of a given return period, say the 1:5 year flood event. Whilst this is a useful measure of categorising the land it does not include a time dependent element; the computational model results give a time history of flood levels throughout the flood season. A better measure of the characteristics of a flood event is therefore the 'depth phase' which gives the percentage of land flooded to different depths throughout the flood season. It is the depth phase which is most useful in determining other factors such as the impact on ecology and fish areas.

POST-PROCESSING ANALYSIS

General

12. The 25 year simulations generate 25 years of daily water levels and discharges at the model nodes. This is a vast amount of information which in its raw form is of limited use, hence there is a need for post-processing. The output required from this post-processing will depend on the use to which it is to be put. The simulation results were used for engineering and fisheries/environment analysis. Figure 1 is a flow diagram illustrating the procedure used.

13. The post-processing involved the following analyses to be undertaken for the different disciplines:

> Engineering : Hydrological analysis to give minimum, mean and maximum values for each decad. In addition, return period values of stage and water level for 2, 5, 10, 20, 50 and 100 year return periods for different durations (1 day, 3 day, etc.) and on a decad basis.

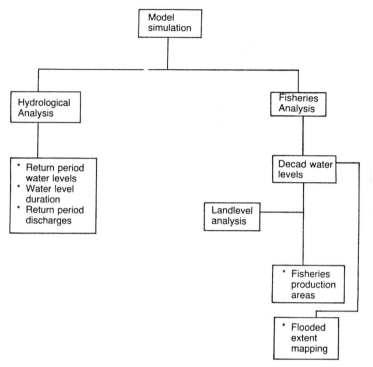

Figure 1 : Post-processing analysis

- Fisheries/Environment : Decad analysis of water levels based on the minimum water level over the decad and return period analysis of flooded areas throughout a hydrological year.

Engineering analysis

14. The engineering analysis is relatively standard and is not discussed in great detail here. The analysis involves the hydrological analysis of water level and discharge time series to give design data for different return periods.

Fisheries analysis

15. For the purposes of this analysis areas which were flooded to a depth greater than 30 cm are potentially suitable for fisheries. The areas flooded to depths greater than 30 cm can be mapped.' This can be done for different return period water levels and different decads by considering the representative water level in the model and the variation in topographic level.

Other analysis

16. Model results can be used to assess the social impacts of proposed engineering measures. The effect floods will have on infrastructure and dwellings can be estimated from model results. In many countries the most significant risks to public health are those from water-related diseases. The model can be used to predict the extent, depth and duration of flooding; this information can be used in turn to assess the likelihood of disease.

CALIBRATION AND VERIFICATION

17. It is crucial in studies in which long time series are used that the model is suitably calibrated for a large range of events ranging from relatively low flow events to the most severest of floods; as this will be the range that is likely to occur within a representative 25 year period.

18. A minimum of 3 flood seasons should be used for the calibration of the model with a further 2 or 3 flood seasons for the verification of the models performance.

19. It must be stressed that the use of the long time series simulations is not to reproduce the historic results but merely to provide a representative series of events on which to base the hydrological analysis. They are a measure of the performance of the hydraulic system under the present conditions and under different developments. In the actual prototype system the topology and topography are likely to be changing with time, as intermediate developments take place, and therefore the response of the system is constantly changing. The long time series runs are undertaken with a fixed topology and topography and would not therefore reflect these changes.

20. In general, the most widely available data for calibrating the models is water levels, and discharges, within the modelled river system itself. Little or no data is available relating to flood levels on the floodplains. However, where this data is available it should be used in the calibration and verification of the model.

21. Other sources of information, non hydraulic, can be used as a verification of a computational models performance. For example, farmers cropping patterns are often dependent on the depth and duration of flooding with different crops being suitable for different hydraulic regimes. By relating model predicted flood depths and durations to potential cropping patterns, the correlation between the model predicted cropping patterns and those obtained from field surveys can serve as a useful further source of verification of the models performance. Indeed, these not only give confidence in the prediction of peak flood levels but also in the rise and fall of the floods throughout the flooding season.

CASE STUDIES FROM BANGLADESH

Flood phase analysis in the North West Region

22. The North West region of Bangladesh is bounded by the Ganges river to the south, the Brahmaputra river to the east, and the Indian border to the north and west, see Figure 2.

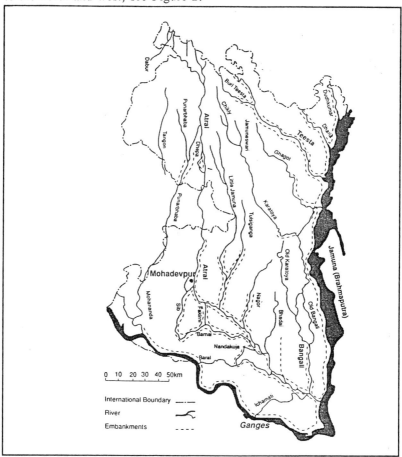

Figure 2 : The North West region of Bangladesh

23. In Bangladesh the flooding situation is commonly classified in flood phases according to depth of flooding as follows:

    F0   0 - 0.3 m
    F1   0.3 - 0.9 m
    F2   0.9 - 1.8 m
    F3   1.8 - 3.6 m
    F4   > 3.6 m

LAND USE CHANGES AND HYDROLOGY

24. Construction of an embankment along the right bank of the Brahmaputra (BRE) was started in the 1960's to prevent spilling from the Brahmaputra into the area. Since its completion, the BRE has suffered continual breaching due to the erosive nature and shifting course of the Brahmaputra. These breaches cause damage to crops and homesteads as they result in rapid rises in water levels on the floodplain beyond those which the rice varieties grown in the area can withstand.

25. In 1991 six main breach sites were in evidence along the BRE. An investigation was undertaken to ascertain the impact of sealing these breaches and the effect this would have on the flood phase categories of the land behind the BRE.

26. The investigations showed that peak flood levels would be reduced by over 1 m along a majority of the area behind the BRE. In an average flood year (1985) the area of land flooded to a depth of less than 0.9 m, F0 + F1, increased from less than 60% to over 90% if the breaches were sealed, Figure 3. In a much more severe flood year (1988) the effect is larger with the percentage figures changing from less than 20% to over 60%. These results indicated that the sealing of the breaches would effectively remove the constraint that flooding imposes on development in the area by allowing a much greater area of land to be planted with high yielding varieties of rice. In addition, the losses due to the sudden breaching of the embankment would be reduced and farmers would be able to plant their crops under a much more predictable flooding regime.

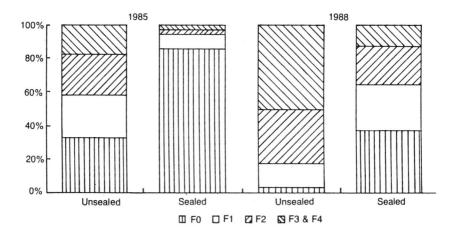

Figure 3 : Change in Flood Phase due to sealing of the BRE breaches

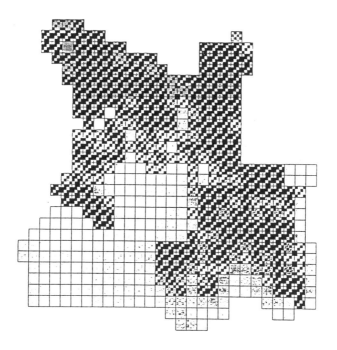

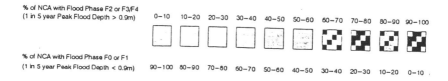

Figure 4 : Present 1 in 5 year peak flood phasing for the Noakhali North Project

Flood phase analysis in the Noakhali North project of the South East region
27. The Noakhali North project lies within the South East region of Bangladesh. It is bounded by the Chandpur Irrigation Project and the Lower Meghna on the west, the Dakatia river in the north, the slight ridge bordering the Little Feni river basin on the east, and the old coastal embankment separating the coastal char lands to the south.
28. Much of the land in the project area is regularly flooded. Only 20 % of the project area is classified as flood free; almost all of this area is in the south west of the project near the Rahmatkhali regulator. The flood problem cannot be considered only in terms of depth of flooding. The timing, rate of rise and duration of flooding are, in addition to the peak attained, very important factors influencing the environment of the region.

LAND USE CHANGES AND HYDROLOGY

Seasonality of flooding is therefore a key variable. In terms of seasonality flooding problems occur in the pre-monsoon period due to the rapid rise in water level, in the monsoon season due to the high flood levels and during the post-monsoon season as a result of congested drainage. In addition tidal flooding can occur as a result of storm surges in the Bay of Bengal.

29. Using the results of the computational model the areal distribution of flooding under present conditions was calculated; this is shown in Figure 4.

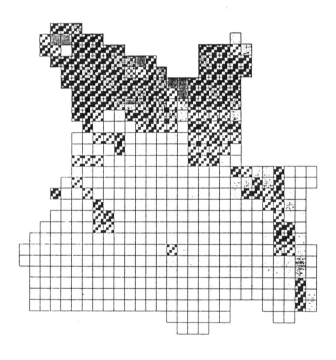

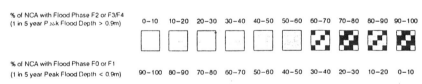

Figure 5 : 1 in 5 year peak flood phasing for the Noakhali North Project with the proposed developments

30. Extensive engineering and hydraulic studies were carried out to identify methods to alleviate the flooding problems in the region. The key components of the final development plan included the

modification of the Rahmatkhali regulator, the installation of a new regulator at Musapur, downstream of the existing Kazithat regulator, together with channel excavation in the khals which feed the Rahmatkhali regulator. The result of these developments on the flooding in the Noakhali North project are presented in Figure 5. Figure 6 shows the changes in flooding which result from the proposed developments.

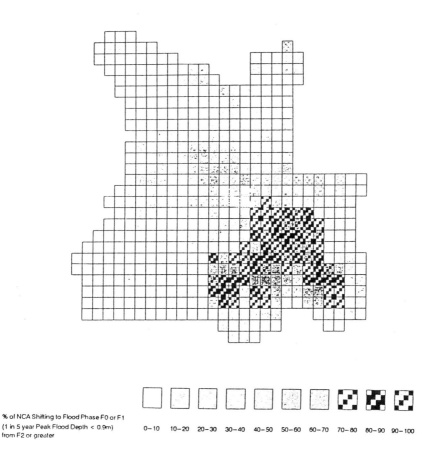

% of NCA Shifting to Flood Phase F0 or F1
(1 in 5 year Peak Flood Depth < 0.9m)
from F2 or greater

0–10 10–20 20–30 30–40 40–50 50–60 60–70 70–80 80–90 90–100

Figure 6 : Change in flood phasing as a result of the proposed developments

CONCLUSIONS AND RECOMMENDATIONS

31. The use of computational models can be expanded by linking the model results to further post-processing software thereby allowing useful information to be available for the many disciplines which are involved in water resources planning.

32. Linking the model results to area/elevation data allows flood depth profiles to be established on the floodplains. Further analysis through the application of environmental or fisheries principles enables potential development scenarios to be established. Together with the pre-project conditions, these can be used to establish the differential benefits and dis-benefits associated with a particular development scenario in many of the disciplines; such as fisheries, environment and sociological impacts.

33. Under pre-project, or present, conditions the post-processing analysis serves as an additional verification of the computational models performance and can give greater confidence in the models ability to predict post-project conditions.

34. The work described in this paper was developed for use in the Flood Action Plan in Bangladesh. The methods employed would benefit from further research in order to refine the methods.

35. Additional work is also required to investigate the applicability of the methods in other flood prone areas in which the flooding regime has a major impact on the environment.

ACKNOWLEDGEMENTS

36. The work described in this paper was carried out by the authors during assignments in Bangladesh. The authors were part of the teams working on the North-West and South-East regional studies. The collaboration and assistance of the other members of the project teams is gratefully acknowledged. The projects were funded by the UK Overseas Development Administration and the World Bank.

Estuarine management: the advantages of an integrated policy analysis approach

B. KARSSEN, M. W. M. KUYPER, M. MARCHAND,
A. ROELFZEMA and M. VIS, Delft Hydraulics

SYNOPSIS. Estuarine management requires a sound understanding of the physical, ecological and socio-economic environment. To support management decisions an integrated policy analysis approach is therefore necessary. This paper presents a framework for such an approach, the applicability of which is illustrated by the Segara Anakan case study (Java, Indonesia).

INTRODUCTION
1. As transition zones between rivers and coastal seas, estuaries feature delicate abiotic and biotic balances, yielding highly productive habitats for various kinds of life. This makes estuarine environments attractive settlements for human activities. Ongoing human activities more and more impose conflicting demands and may deteriorate the natural estuarine environment and most notably in wetland areas. For example, changes in upstream land use, affecting river run-offs, or the amounts of transported sediments, may result in disturbances of the ecological equilibrium by changes of the channel-mudflat system and salt/fresh water gradients, with direct impacts on the habitats for birds, fish etc.
2. This illustrates that for any management strategy of aquatic ecosystems in the watershed, an integrated systems approach is required which takes into account both upstream and downstream relationships. Upstream developments can be considered in the form of changing boundary conditions, which can hence be formulated as scenarios. Often, the impacts on the downstream estuary call for measures in the watershed. Unfortunately, administrative boundaries often do not coincide with the physical boundaries, which makes the implementation of such measures difficult.
3. Clearly, knowledge of the physical and ecological estuarine system is necessary but not sufficient to draft a management or restoration plan. Effectiveness of measures and their social and economic feasibility are all crucial elements in such a plan. Therefore, DELFT HYDRAULICS developed an integrated approach considering the interactions within and between:
- the natural resources system;
- the socio-economic system; and

LAND USE CHANGES AND HYDROLOGY

- the institutional system.

The methodology provides a framework for analysis as a basis for developing appropriate strategies for future use of the water system. DELFT HYDRAULICS, together with ECI consultants applied this integrated approach to the Segara Anakan wetland-lagoon system on Java, Indonesia.

4. The Segara Anakan harbours the only major mangrove wetland left in Java, and is intensively used by local people. It is rich in resources but is rapidly being altered by human activities. A major resource use problem currently exists because of the high levels of sedimentation in the lagoon resulting from changed upland agricultural practices and flood control measures. Added to the increasing pressure on the estuarine environment by local populations, this development threatens the long term viability of the mangrove-lagoon system.

5. An essential part of the applied integrated approach was the combination of mathematical modelling techniques with ecological and socio-economical expertise. A policy analysis approach proved to be an essential framework for this multidisciplinary study.

POLICY ANALYSIS FOR WETLAND MANAGEMENT

6. To support the decision-maker in finding the optimal solution, a policy analysis approach is recommended. This approach can be defined as a systematic process of identifying, analyzing and evaluating alternative courses of action to reach the management objectives. A strong feature of the policy analysis method is the integration of a problem-oriented approach with quantitative analysis by means of modelling techniques.

7. The problem-oriented approach is reflected in the different phases of a policy analysis study (see Figure 1): Inception phase, Development phase and Selection phase (Ref. 1). In the *inception phase*, the subject of the analysis, the objective of the analysis and the constraints are specified. Based on this initial analysis, during which intensive communication with the decision-maker is essential, the approach of the policy analysis is specified.

8. In the *development phase*, the tools for the analysis and possible solutions to the problems are developed. Major activities are data collection and modelling. A preliminary analysis will ensure that the developed tools are suited for the generation and assessment of measures.

9. The preliminary analysis can be turned into a powerful tool in the screening of measures. Testing all kinds of measures, such as engineering structures, dredging operations and upstream land-use improvements often give an insight in the physical behaviour of e.g. the estuary and at the same time give information with regard to the effectiveness of these measures. Thus many measures can be discarded in an early stage of analysis, while other, more promising ones can be combined into a *strategy* and be further analyzed in the next phase.

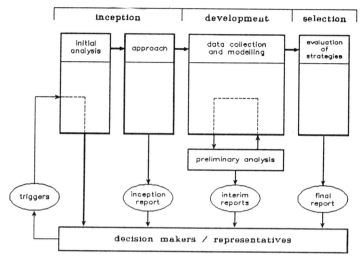

Fig. 1. The policy analysis framework.

10. The purpose of the *selection phase* is to generate a limited number of promising strategies which, after detailed assessment of their effects in terms of the evaluation criteria, are presented to the decision-makers.

CASE STUDY SEGARA ANAKAN

11. The Segara Anakan wetland system is located near the town of Cilacap on the south coast of Java. It contains the largest expanse of mangrove forest left on the island of Java, measuring some 12,000 ha in extent. The mangroves enclose a shallow lagoon of around 1,800 ha. It is subject to tidal action from the Indian Ocean through two inlets. The major inlet is at the southwest corner near where the Citanduy River enters the lagoon. The other main tributary rivers are the Kayu Mati, Cibeureum and Cikonde Rivers. An eastern waterway connects the lagoon to the port of Cilacap through a shallow tidal divide (Figure 2).

12. As early as 1931 it was realized that sedimentation in the lagoon caused the rapid infilling of the lagoon (Ref. 2). In fact, the author of this report drew up the first management plan for the wetland with emphasis on sustainable silviculture. Agricultural mismanagement, deforestation and over-population in the watershed have since then accelerated the infilling rate, eventually leading to the expected disappearance of the wetland-lagoon system by the year of 2000. Most of the sediment load is brought to the lagoon by the Citanduy River and Cikonde River, currently estimated at around $6 \cdot 10^6$ tonnes and $1 \cdot 10^6$ tonnes per year respectively. Figure 3 shows the development of the lagoon from 1903 through the expected geometry in 2000.

13. Since the late forties several study teams have proposed different development plans for the lagoon. Over the years, these plans have evolved from predominantly reclamation

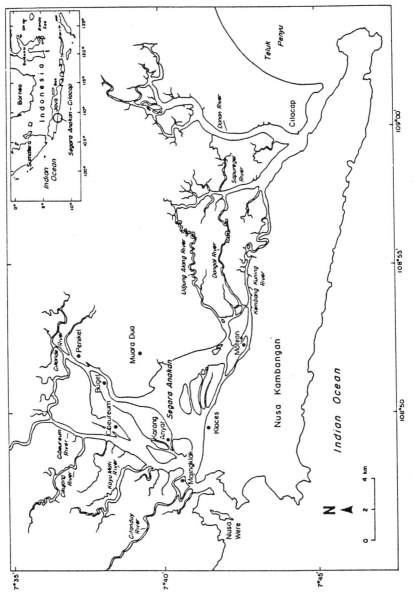

Fig. 2. Situation map of the Segara Anakan.

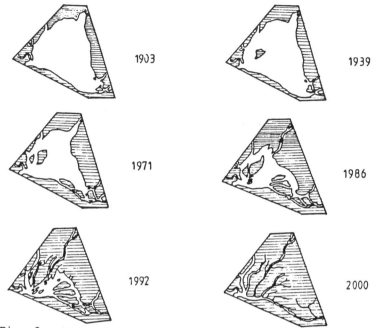

Fig. 3. Maps of the Segara Anakan since the earliest, 1903. The surface area has been decreasing at an accelerating rate. By the year 2000, it is projected that only tidal channels will remain.

oriented to predominantly conservation oriented. One of the main reasons of this change-over in thought was the appreciation of the role of Segara Anakan in the sustenance of the valued lagoonal and offshore fishery yields. Segara Anakan not only provides the livelihood of some 8,000 villagers living within the lagoon but also to the highly profitable shrimp and finfish fishery, currently generating an annual revenue of some ten million US dollars.

14. A continuing sedimentation over the next decades will lead to a complete filling-up of the lagoon area, leaving but a few channels which drain the watershed (Ref. 3). As human pressure around the periphery of the wetland area will only increase, the remaining mangrove forest will largely be converted into low productive rice paddies, even though the soil is highly unsuitable for agriculture because of the high acid-sulphate content (Ref. 4). As a consequence of the disappearance of the wetland, both the artisanal and offshore fisheries productivity are expected to decline drastically.

15. With this in mind, the Asian Development Bank initiated the Segara Anakan Conservation and Development Project - the purpose of which is to assess the technical feasibility and economic viability of various alternative engineering measures to conserve the lagoon-wetland system as an ecological entity and valuable economic resource while preserving existing upstream development. Although soil conservation strategies in

the hinterland would at first sight appear to be the most comprehensible solution, the rate of infilling is such that additional technical measures are necessary in the short term.

16. Most of the aspects of lagoon conservation are so tightly interwoven, that one cannot assess the feasibility of any one engineering option without looking at a large array of consequences, the most obvious of which relate to the economic, environmental, and socio-economic viability. The policy analysis advocated above not only identifies the different consequences but also provides a framework for evaluating and modify the various alternatives. Because of the complexity of the wetland-lagoon system any effort to adequately evaluate the best solution cannot proceed without a major computer modelling exercise. DELFT HYDRAULICS participated in the Segara Anakan Project as subcontractor of ECI for the determination of the technical efficiency of the alternatives considered and for the assessment of the environmental impact of these alternative measures. The related cost, the socio-economic consequences and the institutional aspects were assessed by the contractor ECI.

INITIAL ANALYSIS

17. The engineering measures, found most likely feasible at the start of the project were:

(1) conventional dredging (removal of shoals),
(2) agitation dredging,
(3) enhanced flushing,
(4) construction of a barrage, and
(5) construction of a barrage in addition to the diversion of the Cikonde.

In addition, a
(6) 'do nothing' option (infilling allowed)
was used for comparison. But because there is also a need of maintaining proper drainage channels during the event of high discharge rates a more realistic 'do nothing' option involves
(7) the conventional dredging of drains.

18. Conventional dredging involves the dredging of mud in the centre of the lagoon and disposing it by pumping or hauling away to a suitable area. In this way, the local deepening should serve as a sediment trap, which should lead to smaller sedimentation rates in the whole lagoon.

In contrast, with agitation dredging the mud is stirred up mechanically or hydraulically and it is left for the tidal currents in the lagoon and the density currents induced by the weight of the fluid mud layer itself to dispose of the material.

19. Enhanced flushing is the procedure whereby sea water from the flood tide at Cilacap is directed through the lagoon to the western outlet by means of structures positioned at the tidal divide with the idea that a larger tidal volume will now flow through the western outlet during a tidal period,

resulting in larger flow velocities and thus reduction of the sedimentation.

Both barrage options involve the construction of a barrage across the lagoon near the western outlet thereby bypassing the Cibereum and Citanduy Rivers. The second barrage option moreover diverts the Cikonde. The barrage in both options includes gates enabling animal migration and nutrient influx, and a ship lock.

20. Finally, the conventional dredging of drains implies that the lagoon is allowed to fill up, but maintenance dredging is required in order to maintain the drainage function of the remaining channels.

21. The main criteria in evaluating the different engineering alternatives were:

- cost
  - c1. discounting for Net Present Value, initial costs and annual operating and/or maintenance dredging costs
- technical efficiency
  - c2. how much sediment will leave the lagoon or, in other words, how large will be the sedimentation in the lagoon after taking the proposed measure?
  - c3. how long will it take for the lagoon to fill?
- technical feasibility
  - c4. can the structures be built and operated properly?
- environmental impact
  - c5. is the option environmentally sound?
  - c6. will the fishery yields be sustained?
  - c7. will the mangrove forest survive?
  - c8. will the mudflats be maintained?
- socio-economic impact
  - c9. will the local people be affected?
  - c10. are other socio-economic functions of the lagoon (e.g., transportation) affected?
- policy
  - c11. do the engineering options fit in with existing plans of the local government?

APPROACH

22. In order to answer these questions for each alternative, several physical, environmental, ecological and socio-economical parameters have to be known quantitatively. Moreover, the environmental and socio-economic impact of the alternatives very much depend on the (range of the) values of some physical parameters, such as sedimentation rates, salinities, water levels and flow velocities. Computer modelling is a time- and cost-effective way to provide accurate values of these physical parameters for each of the alternatives considered here under various conditions of the tide (spring, mean or neap) and river discharge (maximum, mean or minimum).

23. As was already mentioned above, the type of problem

involved should determine whether mathematical modelling is necessary or not and if so, to what extent. In this case, very accurate sedimentation rates and salinity distributions were required, whereas the lagoon is characterized by complicated three-dimensional hydrodynamic and (sediment) transport processes. Therefore, it was decided to make use of the state-of-the-art in mathematical modelling: a two- and three dimensional hydrodynamic model (TRISULA), coupled with a quasi three-dimensional sediment transport model (DELMOR).

MODELLING

24. In view of the aforementioned criteria, the following objectives of the modelling exercise were defined:
- determination of the volume of sediments deposited in the lagoon (cf. c1, c2, c3, c7, c8, c10)
- determination of the water levels and flows (cf. c7, c8, c9, c10)
- determination of the salinity ranges (cf. c5, c6, c7)

25. Table 1 gives an overview of the computed volumes of sediments deposited in the lagoon for mean tide/mean discharge conditions and the computed ranges of the salinity in the eastern side of the lagoon (inside the main mangrove area) and in the centre of the lagoon for four of the proposed measures ((1), (3), (4) and (5)) and the present values (8). Figure 4 graphically presents the computed sedimentation pattern in the lagoon for the present situation and mean conditions (adapted from Ref. 5).

Table 1. Sedimentation and salinity ranges for the alternatives (a negative deposition figure implies erosion)

Alternative	Deposition (10^6 m^3/yr)	Salinity range (ppt)	
		Centre of the lagoon	Eastern part of the lagoon
Present situation	2.34	3-12	4-4
Removal of shoals	2.36	3-12	4-4
Enhanced flushing	0.63	8-17	19-24
Barrage	- 1.06	2-2	2-2
Barrage + diversion	- 1.11	26-26	26-26

26. For reasons of technical efficiency and feasibility, the agitation dredging option, measure (2), could be discarded beforehand. This new technology has not been proven in such remote areas as the Segara Anakan and the existing equipment is by far not suitable for this shallow environment. Hence, it would require additional conventional dredging.

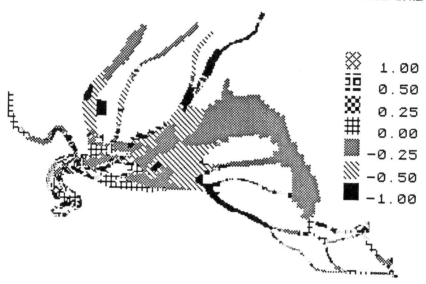

Fig. 4. Sedimentation-erosion pattern in the present situation (negative figures imply net sedimentation).

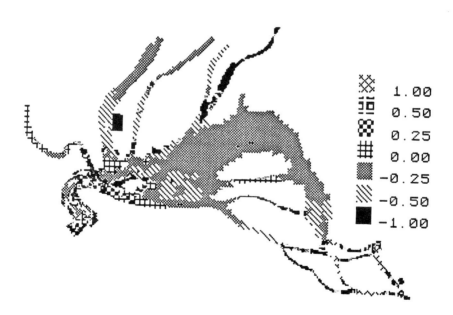

Fig. 5. Sedimentation-erosion pattern with enhanced flushing (negative figures imply net sedimentation).

LAND USE CHANGES AND HYDROLOGY

27. According to this table, the efficiency of both barrage options is clearly the best. However, there are some major drawbacks to both barrage options which render these comparatively costly options unacceptable. Although the technology is proven in other areas of the world, the complexity of the construction and subsequent operation of the barrage hardly warrants technical feasibility. Moreover, it is expected that the gates, although providing an opening to the lagoon, rather pose an obstacle to the migration of fish and shrimp owing to the greatly increased stream velocities and turbidity in the vicinity of the gates. In addition, the reduction in salinity down to brackish conditions does not benefit the mangrove productivity (Ref. 6), nor the nursery function. The commercially important shrimp species encountered in the lagoon grow best at intermediate salinities of about 20-25 ppt.

28. Enhanced flushing also seems promising. The sedimentation is very much reduced, but this is mainly due to a high erosion rate in the main tidal channel of the lagoon. Seeing from Figure 5, it can be inferred that the sedimentation in the shallow mudflat area compares to the present situation, and hence this option is not a guarantee for the preservation of the lagoon. Furthermore, this alternative also causes an increase in sedimentation in Cilacap Harbour, of which the authorities would incur extra maintenance dredging costs.

EVALUATION OF STRATEGIES

29. After weighting the score of each alternative for each of the aforementioned criteria (c1 through c11), it became clear that none of the alternatives that were considered to be feasible beforehand were acceptable.

30. Least unacceptable was the 'do nothing combined with conventional dredging of drains' for reasons of low cost and that the environment is allowed to take its course, the latter because it is perceived natural. The fisheries yield retained with the conversion of the lagoon and mudflat area into mangrove forest will however be considerably less than present values. This is largely due to the overall freshening of the water in the remaining wetland and the considerable reduction of creek, mudflat and lagoon area. The primary habitats of most juveniles are the shallow bays, inlets, creeks and channels and mangrove fringes around the lagoon (Ref. 7) (Figure 6). The mangroves serve largely as the ultimate source of food in the estuarine food chain and their roots further offer protection to larger predators.

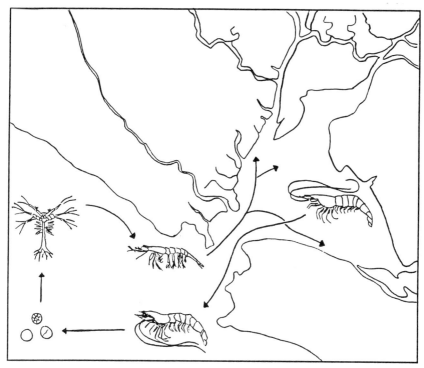

Fig. 6. Generalized life history of shrimp (*Penaeus* spp.) in the Segara Anakan. Shrimp eggs are spawned in the coastal waters. The eggs hatch rapidly and pass through various larval stages until they reach the wetland-lagoon system as post-larvae. During the 2-3 months in the estuary, the shrimp grow rapidly, then migrate back to the Indian Ocean and complete their life cycle (modified from ref. 8).

31. However, during the project, insight and knowledge of the physical and environmental characteristics of the lagoon was gained, which led to the definition of a new alternative:
(8) diversion of the Citanduy in combination with diversion of the Cikonde into the Cibeureum

32. The diversion of the Citanduy is established through excavating an artificial outlet leading direct into the Indian Ocean and closure of the old branch emptying in the lagoon. With this option, it is in theory possible to save the lagoon from the incessant sedimentation and at the same time destroy its productivity as it now exists. The irony is that the sediment that is filling the lagoon carries the very nutrients that make the lagoon so productive.

33. This alternative measure is now being studied in the next phase of the project. Emphasis is put here on the amount of nutrients and sediments which could reach the lagoon from

the newly formed outlet by sea. The diversion of most of the fresh water entering the lagoon might convert the wetland from a river-dominated into a tide-dominated wetland. The consequently higher salinities are conducive to shrimp farming, hence providing an additional source of income to the local people. Moreover, a more saline environment will preclude the conversion of mangrove into agriculture on the periphery of the wetland by local people.

34. Another option which deserves more attention in the ongoing phase of the project is the conventional dredging option (removal of shoals) which with appropriate mitigating measures taken, would preserve the lagoon. Yet, this option is not a feasible option in the long term as it requires a continuing effort of considerable capital dredging. Nonetheless, the calculated net present value (NPV, *i.e.* an economic valuation technique which takes into account present and future coasts and benefits) discounted over a period of 30 years still favour inclusion of this option.

The Segara Anakan project clearly shows how different steps in the policy analysis are taken. The trigger of the study is the expected total infill of the lagoon which unless appropriate measures are taken, will lead to unacceptable environmental, economic and socio-economic impacts. Earlier studies (Ref. 3, 9-11) covered the inception phase and much of the data collection. The consultant ECI (Ref. 3) moreover provided the various alternatives which were to be evaluated in the subsequent development phase on the basis of a set of criteria covering an array of technical, economic, socio-economic and environmental aspects. The results of this phase are largely covered in this paper. One of the alternatives put forward in the inception phase (*i.e.* conventional dredging) and one newly formed alternative are currently being studied in the selection phase. Although initial environmental examination occurred throughout the project, in the final phase, the alternatives are in addition subject to a full Environmental Impact Assessment to be approved by the Government of Indonesia and the ADB. It is further stressed that throughout the project meetings were held with the decision makers to communicate the results of the various project phases and discuss relevant issues.

35. The newly formed alternative shows that a project like this cannot be planned in all details beforehand. In executing the project, it became possible to define a new alternative based on the knowledge and understanding of the natural resource system, the socio-economic system and the institutional system that was built up during the project.

36. Although the project is still in progress, it should be emphasized that the result of any promising alternative being implemented, largely depends on how the local people respond. Their demand of timber and fire wood provided by the mangroves and their need of arable land greatly exceed the natural resources available. For this purpose the engineering measures

should go hand in hand with an appropriate management plan of the wetland area which is to be fully endorsed.

CONCLUSIONS AND LESSONS LEARNED

37. The large number of users of a wetland, or estuary in general, normally gives rise to conflicts. A proper management plan of the wetland is then necessary. The policy analysis advocated here provides a suitable framework for evaluating large-scale interventions.

38. In order to prepare a strategic management plan of a wetland system, (policy) alternatives will have to be defined. These alternatives may lead to changes in the physical system of the wetland/estuary, which in their turn may have socio-economical, ecological and/or political consequences. Therefore, wetland management requires an integrated approach covering ecological, socio-economical, political and institutional aspects.

39. A crude assessment of the environmental and socio-economic impacts of proposed strategies is necessary during the phase of initial analysis of a policy analysis study. In that phase, unrealistic strategies can be dropped from considerations beforehand.

40. The level of accuracy on the information of changes in the physical system that is required for a proper assessment of the ecological, socio-economical, political and institutional impacts of a policy strategy determine whether mathematical modelling should be used as a tool to obtain this information. In case high accuracy is deemed necessary, mathematical modelling is a time and cost effective tool to determine the effects of future changes in a wetland system quantitatively.

41. In order to determine the impact of land use changes on the future geometry of a wetland/estuary, morphological modelling is very important. At this moment, morphological modelling is still in its infancy. Intensive study is needed to develop the knowledge about the physical processes that determine morphological changes.

42. The combination of experts on policy analysis, environmental impacts, socio-economics and mathematical modelling is essential in defining wetland management strategies. It is noted that mathematical modelling as a tool for environmental impact assessments requires a different approach than mathematical modelling for strict engineering purposes.

43. Engineering measures cannot always provide the answers to the problems encountered. In order to achieve an effective solution, it is equally important to develop management plans which cover such areas as public awareness, conservation of mangroves through the control of illegal cutting and conversion, sustainable fisheries management, and upland erosion control. The aim of wetland management should be to find a policy strategy which leads to a good balance between technically efficient measures and socio-economically and institutionally acceptable solutions.

REFERENCES
1. PENNEKAMP, H.A. AND WESSELING, J.W. Methodology for Water Resources Planning. 1993, DELFT HYDRAULICS Report T 635.
2. DE HAAN J.H. Het een en ander over de Tjilatjapsche Vloedbosschen. Tectona, 1931, vol. 24 (1/2), 39-76.
3. ECI. Segara Anakan Engineering Measures Study. Main Report. Citanduy River Basin Development Project, Banjar, 1987.
4. SYUKUR, A. Properties of mangrove soil bordering the Segara Anakan and their significance to land development. Proceedings of the Workshop on Coastal Resources Management in the Cilacap Region, 20-24 August 1980, Gadjah Mada Unviersity, Yogyakarta, 160-169.
5. DELFT HYDRAULICS. Segara Anakan Conservation and Development Project: Mathematical Modelling Study. 1993, DELFT HYDRAULICS Report Z 527.
6. KUYPER M.W.M. Review of the ecological basis for environmental assessment of mangrove ecosystems. 1993, DELFT HYDRAULICS Report T 1153.
7. SASEKUMAR A., CHONG V.C., LEH M.U. AND D'CRUZ R. Mangroves as a habitat for fish and prawns. Hydrobiologia, 1992, vol. 247, 195-207.
8. DAY Jr. J.W., HALL C.A.S., KEMP, W.M. AND YANEZ-ARANCIBIA, A. Estuarine ecology. John Wiley & Sons, New York. 1989.
9. ECOLOGY TEAM. Final task report. Ecological aspect of Segara Anakan in relation to its future management plan. FF/BAU, Bogor, 1984.
10. ECONOMIC TEAM. Final task report. Study of the socio-economic aspect of Segara Anakan. Economic Faculty, Bandung, 1984.
11. ERFTEMEIJER P., VAN BALEN P. and DJUHARSA, E. The importance of Segara Anakan for conservation with special reference to its avifauna. Asian Wetland Bureau/INTERWADER-PHPA, Bogor, 1988, Report No. 6.

Halting and reversing wetland loss and degradation: a geographical perspective on hydrology and land use

G. E. HOLLIS BSc, PhD, MIWEM, Wetland Research Unit, Department of Geography, University College London

SYNOPSIS. Wetlands perform functions that are valuable to human societies. Wetland loss has been enormous. Wetland management must be a wide ranging process extending beyond technical measures in existing wetlands. The aim must be to halt and reverse wetland loss and degradation. The recognition of wetland functions within integrated river basin management and the incorporation of wetland management into territorial and development planning is needed. Wetland restoration, perhaps aided by grants for the reduction of agricultural surpluses, should enhance the water environment. Institutional complexities impede wetland management.

INTRODUCTION

1. Wetland management requires political will to put it into practice, an appropriate institutional structure, an effective legal framework, trained personnel with multi-disciplinary skills, public understanding and support, scientific knowledge of wetland functioning, and the resources, human, technical and financial, to implement the actions decided upon. Wetland management has been notably unsuccessful to date, since there has been an enormous loss and degradation of wetlands worldwide. Whilst data for the last 10 years does not exist, in most countries there is still a steady, and in some cases an accelerating, net loss of wetlands.

2. Wetlands perform valuable functions with significant values for the health, safety and welfare of human societies. Agenda 21 (UNCED, 1992) asserted that "the right to development must be fulfilled so as to equitably meet developmental and environmental needs of present and future generations". The loss of wetlands may be considered as a reduction in the planet's natural capital. Moreover, "some or all (wetland) functions can be restored where they have been lost, or enhanced through rehabilitation measures in degraded wetlands" (IWRB, 1993). Therefore, it is necessary to seek effective means to halt and reverse wetland loss and degradation. It is not sufficient to construe "wetland management" as a set of technical methods to conserve remaining wetland habitats, it ought to be seen as a broad battery of measures to ensure that wetlands play their full role in sustainable development.

3. The geographical perspective encompasses the hydrological cycle, ecosystem functioning, and social, economic and political processes. Within this broad integrative context, the two major themes of hydrology and land use, as determined by the overall structure of this volume of papers, are addressed using a range of case studies.

4. The basis of hydrological science for wetland management is relatively strong, but the application of this knowledge to practical management has to confront a range of problems related to data, information availability, institutional webs and the nature of the decision making processes in many countries. It is argued that the multiple functions and values of present, and past (i.e. restorable), wetlands needs to be recognized within a river basin approach to integrated water management.

5. Pressures from land use change, both directly in the wetland and indirectly in the catchment, reflect the social and economic realities with which wetland management must grapple. Wetland management should be integrated within the territorial planning system. For some wetlands this will involve schemes for the social and economic development of the region surrounding the wetland so as to maximise the opportunities for sustainable development for the wetland and its environs whilst minimising environmental impacts. In other cases, the need to reduce agricultural surpluses etc or to enhance urban habitats could provide opportunities to restore wetlands and to improve the overall health, safety and productivity of the water environment.

DEFINITIONS

6. The internationally accepted definition of a wetland from the Ramsar Convention is: "areas of marsh, fen, peatland or water, whether natural of artificial, permanent or temporary, with water that is static or flowing, fresh, brackish or salt, including areas of marine waters, the depth of which at low tide does not exceed six metres". They may include "riparian and coastal zones adjacent to the wetlands or islands or bodies of marine water deeper than six metres at low tide lying within". This catholic proposition can include rivers, lakes, coastal areas including lagoons and even coral reefs. Mitsch and Gosselink (1993) provide an erudite summary of the problems of wetland definition.

WETLAND FUNCTIONS

7. Whilst the vital functions performed by wetlands have become more fully appreciated (Adamus and Stockwell, 1983), it has been acknowledged that not all wetlands perform all of the functions and not all wetlands perform the functions equally effectively (Table 1). Wetlands provide natural storage areas for flood water reducing the risk of inundation and damage downstream and the retention of water can encourage recharge to underlying aquifers. Wetland vegetation, especially mangroves but also salt marsh to a lesser extent, can stabilise shorelines by reducing the energy of waves and currents and retaining sediment with their roots. Because toxicants (such as pesticides) often adhere to suspended matter, sediment trapping frequently results in

water quality improvements. Water quality can also be improved by the ability of wetlands to strip nutrients (N & P) from water flowing through them. The facility of many wetland plants to pump oxygen to their roots also allows wetlands to treat water polluted with organic material. Mitsch and Gosselink (1993) refer to wetlands both as "the kidneys of the landscape" and as "biological supermarkets" because of the extensive food chain and rich biodiversity they support.

Table 1   Functions and Values of Wetland Types (after Dugan, 1990)

	Estuaries (without mangroves)	Mangroves	Open coasts	Floodplains	Freshwater marshes	Lakes	Peatlands	Swamp forest
Functions								
1. Groundwater recharge	○	○	○	■	■	■	●	●
2. Groundwater discharge	●	●	●	●	■	●	●	■
3. Flood control	●	■	○	■	■	■	●	■
4. Shoreline stabilisation/Erosion control	●	■	●	●	■	○	○	○
5. Sediment/toxicant retention	●	■	●	■	■	■	■	■
6. Nutrient retention	●	■	●	■	■	●	■	■
7. Biomass export	●	■	●	■	●	●	○	●
8. Storm protection/windbreak	●	■	●	○	○	○	○	●
9. Micro-climate stabilisation	○	●	○	●	●	●	○	●
10. Water Transport	●	●	○	●	○	●	○	○
11. Recreation/Tourism	●	●	■	●	●	●	●	●
Products								
1. Forest resources	○	■	○	●	○	○	○	■
2. Wildlife resources	■	●	●	■	■	●	●	●
3. Fisheries	■	■	●	■	■	■	○	●
4. Forage resources	●	●	○	■	■	○	○	○
5. Agricultural resources	○	○	○	■	●	●	●	○
6. Water supply	○	○	○	●	●	■	●	●
Attributes								
1. Biological diversity	■	●	●	■	●	■	●	●
2. Uniqueness to culture/heritage	●	●	●	●	●	●	●	●

Key: ○ = Absent or exceptional; ● = present; ■ = common and important value of that wetland type.

8. For centuries, wetlands have played a central role in the rural economy of many parts of the world, such as Sahelian Africa (eg Hollis et al, 1993c), where floodplains provide highly productive agricultural land, dry season grazing for migrant herds, fish, fuelwood, timber, reeds, medicines and other products in addition to important wildlife habitats. All these functions have both direct economic and other financial values that benefit human welfare, health and safety (Dugan, 1990). The financial valuation of wetland functions and products is advancing rapidly. Barbier et al (1993) have shown that the multiple products of the Nigerian Hadejia-Nguru wetlands that can be easily valued (agricultural crops, grazing and fuelwood) are around six times more valuable per hectare than cropping within the 14,000 ha Kano River Irrigation Project. More importantly still, Barbier et al (1993) showed that returns per unit of natural river inflow or artificially controlled reservoir releases to the wetlands yielded between 242 and 565 Naira/ 000m^3 compared to less than 0.5 Naira/000m^3 in the formal irrigation scheme.

LOSS AND DEGRADATION OF WETLANDS

9. In England, the area of freshwater marsh fell by 52% between 1947 and 1982 and in the early 1980s the rate of drainage continued at between 4,000 and 8,000ha per year. 35% of the ponds around Huntingdon were lost between 1950 and 1969 (Baldock, 1990). In Roman times, 10% of Italy (3 million ha) was wetlands. Only 764,000 ha remained by 1865, and by 1972 this had diminished to only 190,000 ha (Ramsar Bureau 1990). In the Castille-La Mancha region of Spain, 60% of the wetlands have been lost with three quarters of this loss (20,000 ha) taking place in the last 25 years (Montes and Bifani, 1991). In Greece a 60% loss of wetlands, mainly lakes and marshland, took place through two periods of land drainage for agriculture (Handrinos, 1992). In Greek Macedonia for example, since 1930 the area of lakes has been reduced from 586 km^2 to 364 km^2 whilst marshland has declined from 986 km^2 to only 56 km^2 (Psilovikos, 1992). 28% of Tunisian wetlands have disappeared in the last 100 years involving the loss of 15% or 20,854 ha since 1881 (Maamouri and Hughes, 1992). In the lower 48 states of the USA, 53% of the wetlands were lost between the 1780s and 1980s, a loss rate of 60 acres per hour! (Dahl, 1990). In Asia, about 85% of the 947 sites listed in the Directory of Asian Wetlands are under threat with 50% of the threatened sites being under serious threat (Scott and Poole, 1989).

SCIENTIFIC HYDROLOGY IN WETLANDS

10. Hydrology is the driving force and essential common element in wetlands. It is hydrology that puts the "wet" in wetlands! The core hydrological concept of the hydrological cycle functioning within a catchment context does not need elaboration here, save to emphasise its central importance for the measurement, modelling and management of wetlands. As Mitsch and Gosselink (1993) assert "An important point about wetlands ... often missed by ecologists ... is this: Hydrology is probably the single most important determinant for the establishment and maintenance of specific types

of wetlands and wetland processes ... when hydrologic conditions in wetlands change even slightly, the biota may respond with massive changes in species richness and ecosystem productivity. When hydrologic pattern remains similar from year to year, a wetland's structural and functional integrity may persist for many years".

11. Mitsch and Gosselink (1993) (Figure 1) show that a complex hydrological system modifies and determines the chemical and physical properties of the substrate. This, in turn, allows specific biotic ecosystem responses but they and the nature of the substrate feedback to modify the hydrology as the wetland matures and diversifies. The hydrological parameters which act as wetland controls but which can be modified by feedback mechanisms are water level regime, water balance, turnover rate and extremes. Water quality is important, often central, to wetland functioning but it is considered elsewhere in this volume.

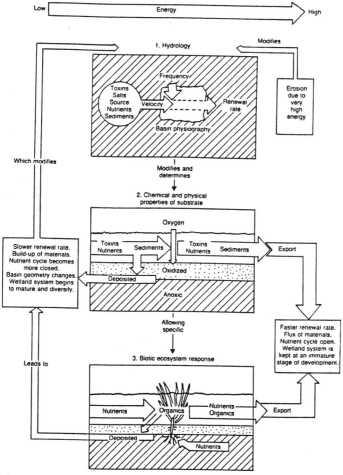

Figure 1   Conceptual model of the direct and indirect effects of hydrology on wetlands (after Mitsch and Gosslink, 1993)

12. The water level regime is the pattern of the water level of a wetland and is a hydrological signature for each wetland type. It reflects the pattern of inflows and outflows, the critical relationships between water depth, flooded area and water volume in the wetland.

13. The water level regime of a wetland is determined by the water balance and the contours of the wetland itself. The water balance of the wetland usually requires attention to the full range of variables including precipitation, river and surface runoff inputs, groundwater inputs and/or outputs, evaporation and evapotranspiration including that from hydrophytes, river runoff from the wetland, any tidal exchanges in coastal sites, and abstraction of water or returns of effluent. The change in water depth in the wetland can be determined from the area, volume and level relationships for the wetland plus soil/substrate characteristics for subsurface water levels.

14. The turnover rate is derived from the division of the volume of water in the wetland by the inflow volume, although in arid regions the outflow volume may give a better indication of turnover. Water residence time is the reciprocal of turnover rate.

15. In many circumstances it is extreme hydrological events, i.e. floods and droughts, which are critical for wetlands. Standard methods of hydrological analysis express the frequency of such events in terms of a recurrence interval. The magnitude of extreme events of a particular recurrence interval is much influenced by the data available. For example estimates of the 100 year flood in the Oued Zeroud above Sebkhet Kelbia in Tunisia changed as a result of a series of high flows from 4,600 m^3/sec in 1964 to 8,000 m^3/sec in 1967, to 13,500 m^3/sec in 1970 and 21,000 m^3/sec in 1975 (Hollis and Kallel, 1986).

REAL-WORLD HYDROLOGY IN WETLAND MANAGEMENT

16. Wetland management requires that society and its institutions modify their actions so as to achieve a set of objectives. Hydrological management is, therefore, more complex and demanding than science based hydrology. The system that controls wetlands as far as management is concerned is not the hydrological cycle but the web of overt and covert institutional linkages. Figure 2 gives an idealised and anonymous representation but such a web exists around all wetlands and potential wetland restoration sites. This web confronts all those seeking to halt and reverse wetland loss and degradation.

17. It is quite normal to find that there is no formal hydrological data for a wetland, be it in the developed or developing world. It is almost always possible to obtain a list of plant species from a library and to find an ornithological group who have counted the birds for at least 10 years but rainfall from a gauge some 10s km away is often the only hydrological data. The legacy of "wetlands as wastelands" has been that hardly any national or regional hydrometric schemes extend into wetlands. The WWF inspired investigation of the impact of groundwater pumping on the Doñana National Park in Spain (Hollis, et al, 1989) concluded that "the present hydrometric network is inadequate for the monitoring, modelling and management of the

aquifer and river systems that support the National Park and the human activities outside the Park. The complete lack of river flow data and of any hydrological or hydrogeological data for the National Park itself are the most serious deficiencies". Investigations into the potential impacts of a new road crossing to the Isle of Sheppey through the North Kent Marshes showed things can be similar in Britain. The marshes have SSSI designation, and international designations under the EU's Birds Directive (79/409) and Ramsar Convention. There was so little formal hydrological data that a special study was commissioned to piece together the available data from disparate formal and informal sources and from established estimation procedures (Hollis et al, 1993a). The study concluded that "there is virtually no hydrological data relating directly to the upland catchments and marshes".

18. In situations where there is data, it may be unavailable because of administrative problems or the reluctance of certain agencies to release it. In the case of the Acheloos River and the associated Messolonghi Ramsar site in western Greece, the river has recently been the focus of a scheme to divert it to Thessalay. One of the issues in the battle over the diversion scheme was the actual flow in the river and the diversion's toll on this. There were flow records since the 1950s available in informal formats for certain sites but the multiplicity of agencies undertaking rain gauging do not make the precipitation data readily available. Hollis (1993b) concluded that "the lack of definitive published hydrometric data for the Acheloos basin impedes analysis. ... the apparent lack of a monograph or catchment review document for the hydrology of the Acheloos is a major handicap to the sound management of the river and its resources".

19. Where a fairly full dataset exists, it is common to find differing interpretations from those intent upon traditional forms of economic development and those striving for wetland management objectives. For example, the Southern Okavango Integrated Water Development Project (SOIWDP) was intended to provide "improved utilization of land and water resources, increased food production, and creation of employment opportunities and a raised standard of living" (Scudder et al, 1993). The Review Team concluded that the flow in the early part of the record, but including the critical design drought period, was over-estimated in the SOIWDP by 20%. "There are major uncertainties in the long term natural outflow from the delta, the extra outflow to be gained from channelization, and losses from the Maun Reservoir". It was considered unlikely that the SOIWDP would have worked as intended" (Scudder et al, 1993).

20. A common hydrological management problem, which derives directly from the institutional power structure controlling a wetland (Figure 2), is that management solutions are sought or demanded, or are only possible institutionally, within the wetland site itself. This may result from the narrowly site-based ecological thinking of nature conservation agencies, from the unwillingness of a water management agency to allocate water to wetland uses, or from powerful (agricultural) forces around the wetland who believe buffer zones, for example, should be in the wetland and not surrounding it.

There has been a long struggle to solve the problems created by the diversion of 65% of the freshwater away from the Ichkeul wetland in Tunisia. It was long asserted that a saving would have to be made on evaporation since no water was available for the wetland from the dams. This lead to a scheme (Hollis, 1986) to divide the wetland with a dyke into a freshwater and a smaller salt water part open to the sea. As such, the "solution" was a scheme resulting in the irreversible loss of 35% of the wetland. In the internationally important wetlands in the Ebro Delta south of Barcelona, wetland management is left largely to the local irrigation cooperatives because the delta is considered to be outside the "basin" of the Hydrographic Commission for the Ebro and, yet, not within the coastal zone managed by the Madrid-based Coastal Management Authority!

21. Whilst models and modelling play such a large role in scientific hydrology, models are relatively much less important in wetland management. This is, in large part, a result of the paucity of data but it reflects another feature of real-world wetland management. Namely, technical inputs can often be of minor importance in water management decisions and related wetland management issues. This is well understood in counties, such as Nigeria, where immediate financial considerations are paramount. The limited value of technical assessments has also recently been fully documented with regard to the Pergau Dam in Malaysia. Here, the House of Commons Committee on Public Accounts (1993) has drawn attention to a host of serious issues relating to the scheme, its financing and its progression through the administrative machine.

22. Whilst almost all hydrologists would take a catchment perspective on their work, there are relatively few wetlands which are managed in the context of their upstream support systems, and the downstream benefits. Indeed, a recent UNDP workshop for south and south-east Asian countries at the Asian Wetland Bureau (1994) concluded that "there is a low level of attention to the hydrological aspects of wetland management which is paralleled by the low level of awareness amongst resource managers of the dependence of wetlands on the maintenance of watershed hydrological regimes". Many relatively dry countries, including for instance Egypt, Tunisia, Morocco and Spain have an explicit, or implicit, policy that "not a drop of freshwater will run to waste in the sea". Whilst this strategy may produce some short term benefits, it is not likely to be sustainable in the long term. Salinization will begin to affect irrigated areas and coastal and other wetlands will be desiccated or so salinized that they will suffer massive ecological change.

23. Agenda 21 (UNCED, 1992) concluded that "the holistic management of freshwater as a finite and vulnerable resource, and the integration of sectoral water plans and programmes within the framework of national economic and social policy are of paramount importance ... coastal area management and development (requires) new approaches that are integrated in content and are precautionary and anticipatory in ambit". Freshwater and coastal wetlands, which are extant or restorable, can make significant contributions to such integrated management approaches once they have been

inventoried, their actual or potential functions assessed and evaluations made of the significance of each of these functions.

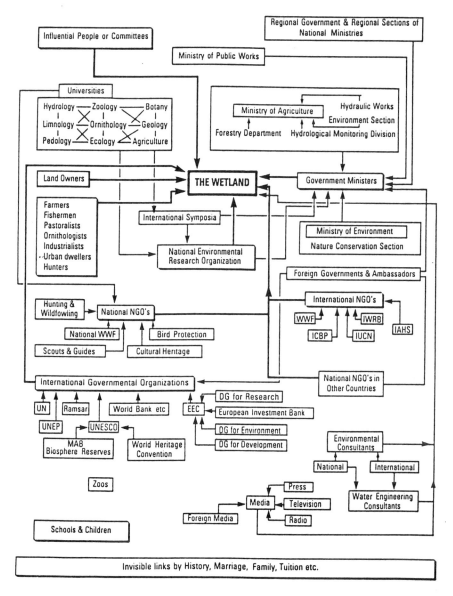

Figure 2   The system controlling wetlands

LAND USE

24. Land use reflects the economic and social development of society (Figure 2). Land use within the catchment influences both the quantity and quality of water feeding wetlands and therefore can affect their ecological character and functioning. Adverse impacts seem to be most common.

Impacts

25. Direct impacts of nutrient enrichment and eutrophication of pools and channels can be found at the British Ramsar sites at Redgrave and South Lopham fen in Suffolk where animal slurry is spread on surrounding fields and at the Swale National Nature Reserve in Kent where the wetland is surrounded on the landward side by intensive arable cultivation. In Spain, groundwater abstraction for irrigation has desiccated the Daimiel National Park in La Mancha with a change from weak artesian conditions to a water table some 30 metres below the surface (Llamas, 1988). Similar groundwater pumping to feed the transformation of scrubby grazing land to strawberry cultivation also threatened the perennially moist ecotone zone of Doñana National Park in SW Spain (Castells et al, 1992). In the Goksu Delta on the Mediterranean coast of southern Turkey, the implementation of a 5500ha irrigation scheme using water from the super-abundant river flow has transformed the Akgol lagoon through the drainage water. A seasonally dry and salty basin has become a rich freshwater habitat with luxuriant vegetation and a remarkable bird fauna that attracts tourists, hunters and international conservation designations (DHKD, 1992).

26. Land use change outside the wetland can have indirect effects on the wetland. Around the Hadejia-Nguru wetlands in Nigeria the long running Sahelian drought has degraded the rain-fed pastures. This has concentrated people and animals in the floodplain wetlands and further stressed their resources. The construction of surfaced roads into the wetlands has further intensified pressure on resources as new marketing opportunities have become available (Hollis et al, 1993c). Fatalities are now common as the sedentary Hausa agriculturalists fight the pastoralist Fulani for access to the fadamas which were once grazed in the dry season but which can now be irrigated with small pumps. Kabaartal Lake in Bihar, India is one of the most important on the sub-continent for birds (Scott, ??). However agricultural intensification in the floodplain lands around have lead politically powerful farmers to demand improved drainage by desiccation of the lake in spite of the 2,000 low caste families who live by fishing the lake. The first steps in the drainage of the lake were undertaken unilaterally by the State Agricultural Service.

27. Land use change within the wetland can fundamentally degrade wetland resources. In the Camargue France, the irrigation of ricefields in the summer maintains large areas of "wetland" habitat and generates large influxes of freshwater to the Etang de Vaccares. However the surface lenses of freshwater below the rice raise the level of the saline groundwater under the surrounding grazing lands leading to a serious salinization of the typical Camargue sansouires with their characteristic horses and bulls (Boulot, 1991). Public or

state ownership of wetlands, which is very common in the Mediterranean, can lead to illegal summer houses as on the Louros Sand Spit near Messolonghi in Greece or around the Goksu Delta in Turkey (DHKD, 1992). Overt subsidy of agriculture has always been a powerful engine of wetland destruction. The subsidized conversion of large parts of the North Kent Marshes to underdrained arable land in the 1970s and 80s was only staunched when significant cash sums were paid for management agreements to safeguard much of the remaining wetland area (Willock, 1993). More recently, Lionhope have (unsuccessfully) proposed major industrial expansion on parts of the North Kent Marshes to ease the unemployment and social blight which affects much of the Swale-Sheerness area.

28. The foregoing review of case studies exemplifies the interlinkages, both hydrological and human, that exist between wetlands, and land use in them and in the surrounding lands. Wetland management must clearly tackle these type of problems and it must therefore involve itself with land use outside the wetlands. In many of the cases cited, the local population in the wetland and environs are relatively poor and disadvantaged. With traditional development models, one of the few opportunities that such people have to improve their lot is to use local resources, ie the wetland, in an unsustainable manner. Having studied these issues around Doñana, the Castells Commission (1992) said "the current development model is unable to sustain stable development, ... and a better quality of life for its inhabitants. Mass tourism is in crisis ... and is not appropriate ... for the extraordinary landscape resources of Doñana ... the irrigation scheme is economically bankrupt, strawberry cultivation is past its peak, and use of the aquifer for rice cultivation is an ecological absurdity".

Planning

29. Therefore, management plans for wetland sites (eg NCC, 1988; Ramsar Bureau, 1993) alone are not likely to be wholly effective in guaranteeing wise use of their resources. There is a need to integrate such plans into the territorial and development planning of the wider region so as to ensure sustainable economic development for the populace in harmony with wetland resource use. Such a process has been launched for Doñana with the Castell's Commission (1992). They concluded that their "assessment reveals the possibilities for ensuring a dignified way of life for the inhabitants of the area, based on rational use of its tourism and natural resources, as well as a programme of public infrastructure investments and the improvement of the skill levels in the area's population". BECOM (1993) have been contracted by the Tunisian Government, using German aid funding, to produce "an ecological management plan (for Ichkeul) taking account of developments already realized or deemed necessary and including institutional measures and laws needed to put the plan into operation plus a programme of social and economic development for the area surrounding the Park". For the North Kent Marshes a major funding request has been submitted to develop a major ecotourism operation based around the marshes but bringing valuable

Enhancement Measures

30. The reversal of the downward trend of wetland area could be reversed if land use planners and water managers collaborated to plan land for the restoration of wetlands. There is now sufficient experience in wetland restoration to suggest that it is feasible to restore some or all wetland functions to the site of a "lost" wetland (IWRB, 1993). The experience with wetland creation at previously dry sites is more equivocal but it is clear that for both restoration and creation to be successful the hydrology of the site must be satisfactory (Kusler and Kentula, 1990). A major change is underway in agriculture in the EU, and elsewhere, as a result of changed policies on subsidies. Large sums of money have been allocated to effect the transformation through set-aside, farm woodlands, environmentally sensitive areas etc. This suggests that there is a real opportunity for wetland restoration on agricultural lands requiring substantial expenditure on drainage or flood defence. In urban areas, the restoration of rivers and wetlands could provide a major boost to wildlife habitats, environmental quality and amenity. The restoration of the environmental services provided by wetlands to rivers may do a lot to offset current problems and to enhance the overall water environment. Three brief examples illustrate the point.

31. In SW Sweden, traditional water meadows are being re-established to strip nutrients from river water to reduce the algal blooms which afflict the coastal zone of the Kattegat. It has been shown that the marginal economic cost of removing 1 kg of nitrate nitrogen by the water meadows was 67 cents compared to $33 by a conventional treatment works (Eriksson, 1990).

32. In S London, an artificial wetland (0.7 ha) has been dug in a school playing field to attenuate the rapid runoff from a shopping complex. A conservation initiative planted the site with reeds and created a deep water pond to bring wetland fauna and flora back to the area and to provide an educational resource. This "green filter" has been shown to strip the pollutants from the stormwater before it enters the Pyl Brook. Cutbill (1993) found that, on the basis of weekly sampling of the inflow and outflow during the 1991, dissolved oxygen increased by 4.6% and sodium and potassium levels increased marginally. The important pollutants all declined; phosphate by 35.9%, nitrite by 20%, nitrate by 68.9%, turbidity by 55.8%, faecal coliforms by 77.5%, copper by 71.4%, zinc by 36.8%, lead by 6.8% and cadmium by 5.9%.

33. The River Quaggy (18.4 km^2) in SE London is heavily urbanised. Severe flooding of property lead to a proposal to further channelise the river with extensive use of concrete. A citizen's group argued strongly for a storage solution and identified numerous sites that could store flood water on open land. This also provided opportunities for improved low flow, water quality, wildlife, river fisheries, amenity and educational sites through small scale wetland restoration at some of the sites. Table 2 shows that the storage scheme was attractive financially.

Table 2 River Quaggy flood alleviation options (National Rivers Authority, 1993)

Scheme	Downstream flow (m³/sec)	Cost (£ million)	Cost/Benefit Ratio
Original channelisation scheme	20.2	6.23	1.08
All feasible storage sites	15.8	4.92	1.36
Largest storage site with some channelisation	17.5	6.00	1.12

34. This more environmentally friendly scheme for the Quaggy is in line with the London Planning Advisory Committee's (1993) view that London's rivers are one of "its greatest but most undervalued resources ... and all proposals for development must ... safeguard the ecological and conservation value of the water, banks and associated open spaces".

CONCLUSION

35. The management of wetlands demands political will, effective institutions and inter-institutional mechanisms, a proper legal framework, a multi-disciplinary staff, public support, scientific knowledge, and the means to implement decisions. Wetland management has not yet stemmed the enormous loss and degradation of wetlands worldwide. Whilst the dynamic nature of wetlands make precise definition difficult, it is now widely accepted that they can perform a range of functions and produce numerous products, which are valuable to human societies. Recent economic studies have shown wetlands to outperform formal irrigation schemes several times over.

36. Hydrology is probably the single most important determinant for the establishment and maintenance of specific types of wetlands and wetland processes. When hydrologic conditions in wetlands change even slightly, the biota may respond with massive changes in species richness and ecosystem productivity. Water level regime, water balance, turnover rate and extremes are the most important hydrological parameters aside from water quality which is covered elsewhere in this volume. The application of hydrological science to practical wetland management confronts problems related to the minimal hydrometric networks in most wetlands, information availability, institutional webs and the nature of the decision making processes in many countries. The multiple functions and values of present, and past (i.e. restorable), wetlands needs to be recognized within a river basin approach to integrated water management.

37. There is usually a complex web of overt and covert linkages between institutions, agencies and individuals that relate to a particular wetland. Many sectors of activity relate to wetlands and yet there are few organisations that exist which can integrate their approaches. Land use affects wetlands in

innumerable ways, directly and indirectly from the lands around the wetland and often very directly by land use change in the wetland. The management of wetlands should be integrated within the land use planning system. For some wetlands this will involve schemes for the social and economic development of the region surrounding the wetland so as to maximise the opportunities for sustainable development for the wetland and its environs whilst minimising environmental impacts. In other cases, the need to reduce agricultural surpluses etc or to enhance urban habitats could provide opportunities to restore wetlands and to improve the overall health, safety and productivity of the water environment.

BIBLIOGRAPHY

Adamus, P. and Stockwell, L. 1983 *A method for wetland functional assessment*. Vols I and II. Reports FHWA-IP-82-23 and 24 US Department of Transportation, Federal Highway Administration, Washington. 181 and 134pp.

Asian Wetland Bureau 1994 *The UNEP/AWB Scoping Workshop on Asian Wetlands in relation to their role in watershed management*. Internal Report of Asian Wetland Bureau, Kuala Lumpur, Malaysia.

Baldock, D. 1990 *Agriculture and Habitat Loss in Europe*. World Wide Fund for Nature and European Environmental Bureau, London. 60pp.

Barbier, E., Adams, W.M. and Kimmage, K. 1993 Economic valuation of wetland benefits. In: *The Hadejia-Nguru Wetlands: environment economy and sustainable development of a Sahelian floodplain wetland*, Hollis, G.E., Adams, W.M. & Aminu-Kano, M. (Eds), IUCN Gland Switzerland and Cambridge, UK. pp191-210.

BCEOM 1993 Etude pour la sauvegarde du Parc National de l'Ichkeul: Rapport de Demarrage. BCEOM, Montpellier, France. 58pp.

Boulot, S. 1991 *Essai sur la Camargue: Environnement, Etat des lieux et Prospective*. Actes Sud, Arles. 89pp.

Committee of Public Accounts 1994 Seventeenth Report: Pergau Hydro-electric Project. House of Commons, London. HMSO. 29pp.

Cutbill, L.B. 1993 Urban stormwater treatment by artificial wetlands: A case study. In: *Proceedings of the 6th International conference on Urban Storm Drainage*, Niagara Falls, Canada. Volume II, p1068-1073.

Dahl, T. 1990 *Wetland losses in the United States 1780s to 1980s*. US Department of the Interior, Fish and Wildlife Service, Washington, 21pp.

DHKD (Dogal Hayati Koruma Dernegi) 1992 *Towards integrated management in the Goksu Delta: A Protected Special Area*. DHKD, Istanbul. 272pp.

Dugan, P.J. (Ed.) 1990 *Wetland Conservation: Review of Current Issues and Required Action*. IUCN-World Conservation Union Publication, Gland, Switzerland, 96pp.

Eriksson, S. 1990 Re-creation of wetlands for nitrogen retention. In: Ramsar Bureau, *Proceedings of the fourth Conference of the Contracting Parties to the Ramsar Convention*, Montreux, Switzerland, Volume II Conference

Workshops. Ramsar Bureau, Gland, Switzerland, p 178-180.

Handrinos, G. 1992 Wetland loss and wintering waterfowl in Greece during the 20th Century: a first approach. In: Finlayson, M., Hollis, G.E. and Davis, T. (Eds.) *Managing Mediterranean Wetlands and their Birds*, IWRB Special Publication 20, Slimbridge, England, p183-187.

Hollis, G.E. 1986 The modelling and management of the internationally important wetland at Garaet El Ichkeul, Tunisia. *International Waterfowl Research Bureau Special Publication No. 4.*, IWRB, Slimbridge, 121pp.

Hollis, G.E. and Kallel, R. 1986 Modelling the effects of natural and man induced changes on the hydrology of Sebkhet Kelbia, Tunisia. *Trans. Inst. British Geographers*, 11 (1), pp86-104.

Hollis, G.E., Mercer, J. and Heurteaux, P. 1989 *The implications of groundwater extraction for the long term future of the Doñana National Park, Spain.* Report to Worldwide Fund for Nature (International). University College London, Dept of Geography. 53pp and Appendix 9pp.

Hollis, G.E., Fennessy, S. and Thompson, J. 1993a A249 Iwade to Queenborough: Wetland Hydrology. Final Report to Ove Arup and Partners, London. 168pp.

Hollis, G.E. 1993b Η ΥΔΡΟΛΟΓΙΚΗ ΘΕΩΡΗΣΗ ΤΗΣ ΕΚΤΡΟΠΗΣ ΤΟΥ ΑΧΕΛΩΟΥ. In: *ΕΚΤΡΟΠΗ ΤΟΥ ΑΧΕΛΩΟΥ: ΣΤΟΙΧΕΙΑ ΑΝΤΙΚΕΙΜΕΝΙΚΗΣ ΕΚΤΙΜΗΣΗΣ*. Hellenic Ornithological Society, WWF Greece, Eliniki Eteria, Hellenic Society for the Protection of Nature, Athens. pp27-56.

Hollis, G.E., Adams, W.M. & Aminu-Kano, M. (Eds) 1993c *The Hadejia-Nguru Wetlands: environment economy and sustainable development of a Sahelian floodplain wetland*, IUCN Gland Switzerland and Cambridge, UK. 244 pp.

Castells et al - International Commission of Experts 1992 *Strategies for the sustainable economic development of the Doñana region*. Translation by Wetland Research Unit, Department of Geography, University College London. 93pp.

IWRB (Int. Waterfowl and Wetlands Research Bureau) 1993 Conclusions from Workshop on Restoring and rehabilitating degraded wetlands. In: *Waterfowl and Wetland Conservation in the 1990s - A Global Perspective*, Moser, M., Prentice, R.C. and van Vessem, J. (Eds.) Proceedings of an IWRB Symposium, St. Petersburgh Beach, Florida, November 1992. pp222-223.

Kusler, J. and Kentula, M.E. 1990 Wetland Creation and Restoration: The status of the science. Island Press, Washington, D.C. 594pp.

Llamas, M.R. 1988 Conflicts between Wetland Conservation and Groundwater Exploitation, *Environmental Geology*, 11(3), pp 241-251.

London Planning Advisory Committee 1993 *Draft 1993 Advice on Strategic Planning Guidance for London: For Consultation*, June 1993. LPAC, London.

Maamouri, F. and Hughes, J. 1992 Prospects for wetlands and waterfowl in

Tunisia. In: Finalyson, M., Hollis, G.E. and Davis, T. (Eds.) *Managing Mediterranean Wetlands and their Birds*, IWRB Special Publication 20, Slimbridge, England, p47-52.

Mitsch, W.J. and Gosselink, J.G 1993 *Wetlands*. 2nd Edition. Reinhold, New York. 722pp.

Montes, C. and Bifani, P. 1991 Spain. In: Turner, K. and Jones, T. (Eds.) *Wetlands: Market and Intervention Failures - Four Case Studies*. Earthscan, London. p144-195.

National Rivers Authority 1993 *Quaggy River Project Liaison Group: 17th June 1993*. National Rivers Authority, London. Mimeo.

NCC (Nature Conservancy Council) 1988 *Site Management Plans for Nature Conservation: a working guide*. NCC, Peterborough. 40pp.

Psilovikos, A. 1992 Prospects for wetland and waterfowl in Greece. In: Finalyson, M., Hollis, G.E. and Davis, T. (Eds.) *Managing Mediterranean Wetlands and their Birds*, IWRB Special Publication 20, Slimbridge, England, p53-55.

Ramsar Bureau 1990 *Directory of Wetlands of International Importance*. Ramsar Convention Bureau, Gland, Switzerland. 782pp.

Ramsar Bureau 1993 *Guidelines on management planning for Ramsar Sites and other wetlands*. Annex to Resolution C.5.7 Fifth Meeting of the Conference of the Contracting Parties to the Convention on Wetlands of International Importance Especially as Waterfowl Habitat, Kushiro, Japan. June 1993. Ramsar Bureau, Gland, Switzerland. Volume 1 pp186-192.

Scott, D.A. 1989 *A directory of Asian Wetlands*. IUCN, Gland, Switzerland and Cambridge, UK. 1181pp.

Scott, D. and Poole, C. 1989 *A status overview of Asian Wetlands*. Asian Wetland Bureau Publication 53, Kuala Lumpur, Malaysia. 140pp.

Scudder, T., Manley, R.E. et al 1993 *The IUCN Review of the Southern Okavango Integrated Water Development Project*. IUCN, Gland, Switzerland. 544pp.

UNCED 1992 Earth Summit '92. Rio de Janeiro 1992. Regency Press, London. 240pp.

Willock, C. 1993 Farming wildfowl on Elmley Island. *The Field*, February 1993, pp82-83.

The value of regional perspective to coastal wetland restoration design

Dr A. M. BARNETT, President, MEC Analytical Systems Inc., R. S. GROVE, Senior Research Scientist, Southern California Edison Company, and S. L. BACZKOWSKI, Wetland Scientist, MEC Analytical Systems, Inc.

1. The restoration of coastal wetlands has become a critical objective of ecologists, resource managers, and the public concerned about this valuable yet dwindling resource. Because of the need to improve habitat and better manage degraded wetlands, the relatively new science of wetlands restoration is rapidly gaining application as a trade-off or mitigation to associated development.

2. Most of the past wetland restoration projects on the west coast of the United States have focused on engineering criteria rather than ecological criteria (Josselyn and Buchholz, 1982). Biological considerations and local and regional ecological requirements have only recently been evaluated in designing restoration projects. This paper describes the value of a regional perspective in wetland restoration and the manner by which results of a regional analysis were incorporated into the restoration designs of San Dieguito Lagoon, San Diego County, in southern California (Fig. 1).

3. Sorensen (1982) and Perl (1990) identified five reasons for conducting a regional analysis prior to wetland restoration design. These are to:
- Understand the distribution and amount of wetland habitat and thus determine their scarcity and the contribution of their creation;
- Allow a comparative evaluation to assess the type and mix of habitats that are optimal for biodiversity or target species;
- Assess the function of migratory bird and fish support; and
- Establish a regional baseline against which to measure improvements in a restoration project.
- Evaluate the potential for increased resilience of populations, especially target species, resulting from the planned restoration.

4. In addition to these, we have found one other advantage as a result of our study. A regional perspective improved the public's understanding of the biological tradeoffs that must occur in wetland restoration planning.

5. Finally, it should be pointed out that once these studies are complete, the information can be kept current and applied to other wetland restorations in the same region.

HABITAT SCARCITY AND COMPARATIVE DISTRIBUTION
6. Southern California marine wetlands are contained between ridges where seasonal rivers or streams flow approximately perpendicular to the shoreline. As

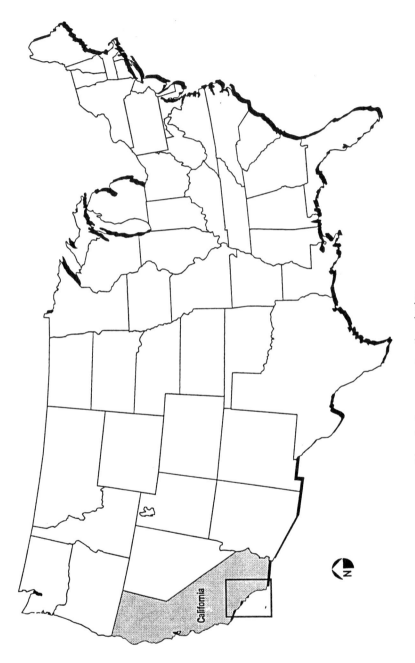

Fig. 1. Regional considered in San Dieguito Lagoon restoration design

freshwater rivers approach their ocean outlets, their waters spread or fan out through the low lying areas behind sand berms of the dynamic beach and littoral zone of the shoreline.

7. Information was gathered from the 16 wetlands between the southern reaches of the Los Angeles, California, metropolitan area and the United States/Mexican border (Fig. 2). The distribution and areas of coastal marine habitat types were obtained from color aerial photographs at 1:500 scale, taken between January and March 1993. Information on topographic relief and surface hydrology from the most recent maps available from the United States Geologic Service and the National Wetland Inventory was used to substantiate photographic interpretation. Habitat types for the Bolsa Chica wetland were determined from information provided in a recent environmental impact report (Chambers, 1992).

8. The habitat types from open water, subtidal areas to uplands (within the 10 foot mean sea level contour) were delineated and summarized in a histogram as acreages for each wetland (Fig. 3). From the histogram it can be seen that open water, subtidal areas are the most prevalent habitat among the wetlands and that they are fairly well distributed across the region. Therefore, it can be concluded that, based on this regional distribution, additional subtidal habitat is not a high priority for the San Dieguito Lagoon restoration.

9. Permanent, vegetated salt marsh habitat and tidal mud flats are each somewhat less in acreage than open water habitat. Although these two habitat types are distributed throughout the region, they are most extensive in the north and south (Fig. 3). Seasonal salt marsh habitat is scarce. Extensive seasonal salt marsh acreage is found only to the north (Bolsa Chica) and south (Tijuana Estuary), although some exists at San Dieguito Lagoon. Brackish/freshwater marsh exists primarily at San Elijo and Buena Vista Lagoons, both of which are in the middle of the region near San Dieguito Lagoon (Fig. 3).

10. Wetlands with permanently open entrances to the ocean tend to have higher biodiversity, better water quality, and more vegetation. Open lagoons also have great value as nursery areas for many marine fishes, such as the California halibut, an important sports and commercial species (Kramer, 1990), and atherinids, a forage species for the California Least Tern (Zedler 1982, PERL, 1990). Those wetlands with entrances continuously open for the last 20 years or longer are identified in Fig. 3. Other than Agua Hedionda Lagoon, these are found at the northern and southern reaches of the region. Thus, an open entrance at San Dieguito Lagoon would be desirable from a distributional perspective.

11. Due to, (1) their relative absence at San Dieguito Lagoon, and (2) regional acreage size and distribution criteria, restoration of vegetated marsh and mud flats is somewhat favored over subtidal habitat. Because of the proximity of brackish marshes in wetlands just to the north of San Dieguito Lagoon, additional brackish habitat at San Dieguito Lagoon would contribute less to the regional picture than would additional seasonal salt marsh there. The areas of seasonal salt marsh at San Dieguito Lagoon should be retained and protected because of their scarcity and the ecological value of this habitat for bird feeding and resting. However, there

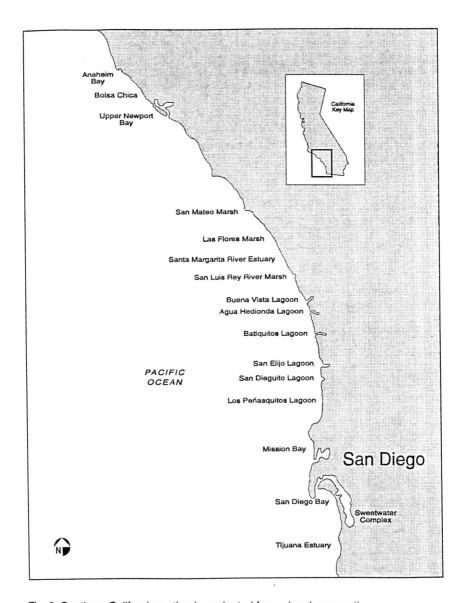

Fig. 2. Southern California wetlands evaluated for regional perspective

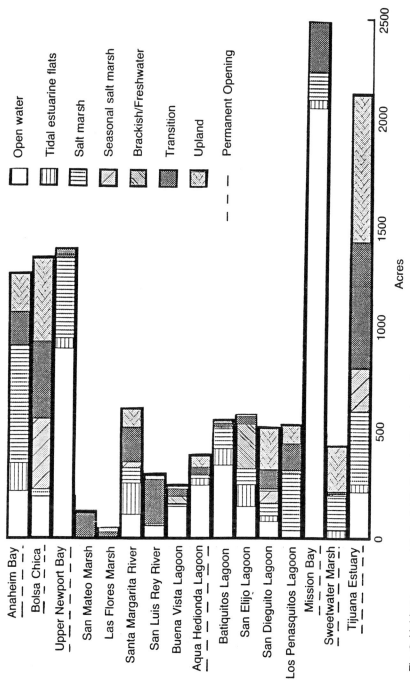

Fig. 3. Habitat types and their distributions in 16 southern California wetlands

may be less advantage to increasing seasonal salt marsh acreage at San Dieguito Lagoon than to establishing new acreages elsewhere, again, based on the distribution perspective of Fig. 3.

HABITAT MIX, BIODIVERSITY, AND TARGET SPECIES

12. A regional analysis can help in the determination of the type and mix of habitat(s) that are optimal for biodiversity or target species. Biodiversity of wetlands is enhanced by a mix of different habitats. For most southern California wetlands, the greater the number and distribution of habitats, the greater the number and variety of species and the greater the interactions among the different species. If one habitat is favored and dominates the wetland to an extreme, it does so at the expense of diversity. For example, increased open water, subtidal habitat will enhance fish and benthic populations but limits bird forage and nesting sites, while an addition of vegetated salt marsh and mudflat areas will tend to increase bird and invertebrate usage.

13. When habitat for a target species is desired, we can examine all the wetlands where that species is successful and unsuccessful in the region and estimate the optimal combinations of vegetation, hydrology, and support species necessary for its success. Cordgrass can be used as an example of a vegetation type that determines the distribution of an endangered species, the Light-footed Clapper Rail. Thus, understanding the factors of cordgrass success is imperative to creating suitable clapper rail habitat.

14. Diversity also is enhanced in wetlands that have refuge areas. During catastrophic events such as flooding, fauna have a much greater chance of survival if there are protected areas where they can take refuge. For example, shorebirds are known to move up into transition and upland zones during periods of very high water.

15. Another significant feature of wetlands with high species diversity is a permanently open ocean entrance. Regular tidal flushing and water renewal will flush the mudflat areas and greatly increase the invertebrate and smaller fish populations and create increased forage area for birds and larger fish.

REGIONAL MIGRATORY ANALYSIS

16. The southern California regional wetlands are a significant portion of the Pacific Flyway, the primary bird migration route along the eastern Pacific Ocean. With the loss of wetland acreage, it has become increasingly important to reestablish appropriate resting and feeding areas of migrating birds. Analysis of wetland stopover sites has shown that preference is related to the sum of seasonal, brackish, and mudflat areas. The greater the acreage of these habitat types, the greater the usage by migratory birds. All of the wetlands designated as major Pacific Flyway wetlands have a moderately high acreages of combined mudflat, brackish, and seasonal habitat (Fig. 4).

17. In the case of the California Least Tern, which migrates from Central America to southern California each spring, habitat preference includes large expanses of bare or sparsely vegetated substrate, a low level of disturbance, and access to

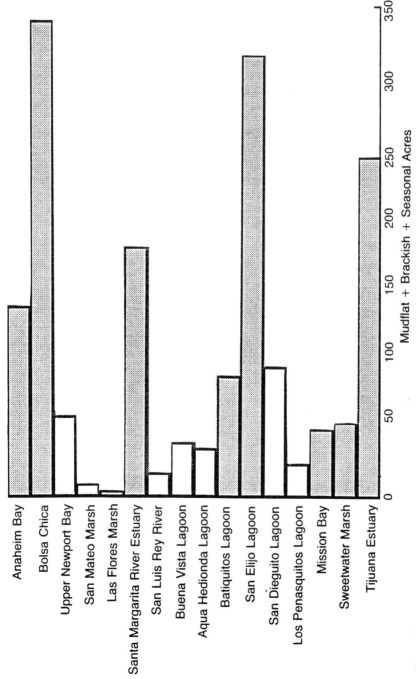

Fig. 4. Sums of acreages of mudflat, brackish and seasonal salt marsh habitats. Acreage of wetland identified as major Pacific flyway stopovers for migratory birds are shaded

bodies of shallow water. The number of pairs are greatest in Anaheim Bay, Bolsa Chica, Santa Margarita River, Mission Bay, and San Diego Bay (Fig. 5).

18. An open entrance to the ocean (tidal flushing) does not seem to be a significant factor in selection of a resting/feeding site by migratory birds from a regional perspective. However, it does affect the preference of the California Least Tern.

19. Using migration as a criterion would suggest that the San Dieguito Lagoon restoration should focus more on mudflat areas and less on vegetated midlevel and high level marsh. Additionally, there is good potential for expansion of the seasonal marsh at the site since that habitat already exists there.

SPECIES RESILIENCE AND WETLAND LINKAGE

20. Catastrophic events such as flooding, entrance closure, or human disturbance periodically occur in wetlands. These events can affect many crucial aspects of the ecosystem. Food supplies may be interrupted, salinities may change drastically, habitat may be altered or destroyed, and species may be eliminated from a wetland. Generally, southern California wetlands have a low level of resilience. Recovery by sensitive species can be difficult and slow and is dependent upon the strength of the gene pool, the nearness to other habitats, and the presence of aquatic and terrestrial corridors.

21. Currently, the overall gene pool of wetland flora and fauna is reduced due to the degradation and loss of wetland acreage. For example, viable breeding populations of the Light-footed Clapper Rail are limited to two wetlands: Upper Newport Bay and Tijuana Estuary (Fig. 6). The Light-footed Clapper Rail is associated with cordgrass, is a poor flyer, and has reduced capability to recolonize a wetland once it has been eliminated. Distance from viable breeding populations and the presence of terrestrial corridors become major factors in clapper rail recruitment and recovery within a given area. In contrast, the Belding's Savannah Sparrow (Fig. 7) has a relatively good, widely dispersed gene pool compared to other threatened species. To maximize Belding's Savannah Sparrow resilience at San Dieguito, the addition of pickleweed can expand the population that is already relatively strong compared to other threatened species.

22. Recolonization by plants also requires proximity to a source population when an entire population has been eliminated. Propagules, or seeds of wetland plants, must come from a nearby area, and repopulation usually occurs by passage through the lagoon entrance. In these situations, refuges from floods are necessary to maintain the regional gene pool, particularly with species that do not establish persistent seed banks (e.g., *Salicornia bigelovii*) (Perl, 1990). Similarly, insect colonization also is dependent upon nearness to other populations due to limited migratory capabilities of most insect species.

23. Resiliency of fish and invertebrate populations requires pelagic connections and access to larvae of other wetlands. A permanently open lagoon entrance is important at San Dieguito because many fish and invertebrates have various life stages that require a wetland habitat with an open connection to the ocean. Since

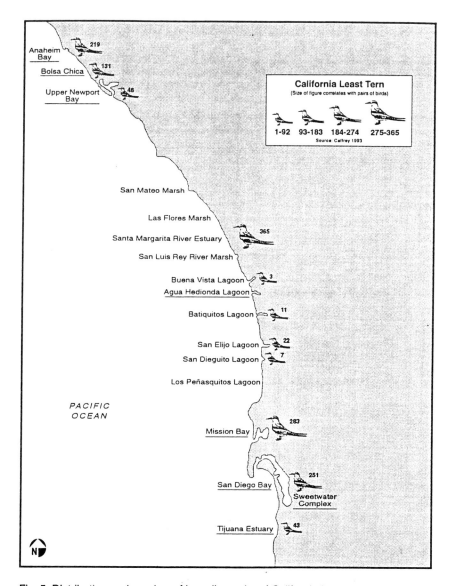

Fig. 5. Distribution and number of breeding pairs of California Least Terns in southern California wetlands

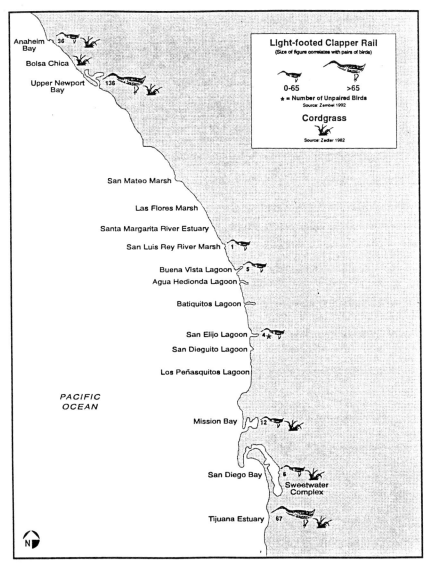

Fig. 6. Distribution and number of breeding pairs of Light-footed Clapper Rails in southern California wetlands

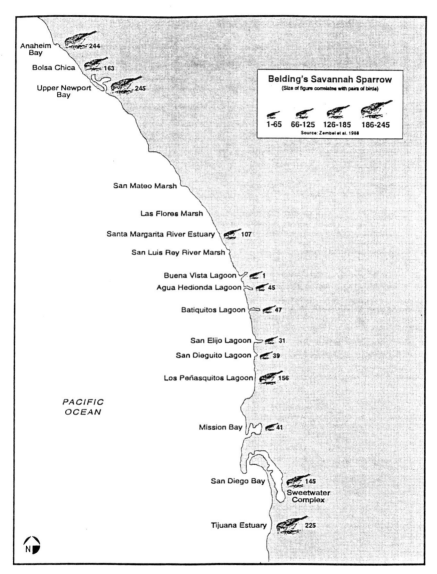

Fig. 7. Distribution and number of breeding pairs of Belding's Savannah Sparrows in southern California wetlands

the effects of flooding can vary greatly among wetlands, recruitment from nearby wetlands via pelagic connections may be necessary.

24. In wetlands with the potential for catastrophic events, a good degree of connection with other, less susceptible wetlands is important. Vegetated terrestrial corridors and stopovers (*sensu* Gibbs, 1993) are essential for species such as the Light-footed Clapper Rail. Cordgrass has never been known to cover significant areas at San Dieguito Lagoon, but presently there are small, experimental patches of planted cordgrass (MEC, 1993). Though the beds are too small to accommodate a clapper rail population, increasing the amount of cordgrass area could increase the likelihood that this wetland will become an important link among wetlands for transient clapper rails. The issue of natural corridors for other species, including mammals, needs further study and development.

BASELINE COMPARISONS AND RESTORATION EXPECTATIONS

25. Establishing a regional quantitative baseline for wetland vegetation and flora is important in order to establish a reasonable wetland profile. Southern California wetlands are spatially heterogeneous and exhibit interannual variability. Baseline data could eliminate unrealistic expectations of a restoration project. Since a given wetland is unlikely to provide optimal habitat for all wetland species, it is necessary to understand the variation in "success" that can be expected from various combinations of habitats. For example, Los Peñasquitos Lagoon, just to the south of San Dieguito Lagoon, provides midmarsh habitat that supports lush stands of pickleweed and Belding's Savannah Sparrow but poor migratory and shorebird populations, while San Elijo Lagoon, adjacent to San Dieguito Lagoon to the north, has a large brackish water body with mudflats that support large numbers of migratory and shorebirds. The ocean entrances of both of the adjacent wetlands are closed most of the time so both have relatively depauperate marine wetland and nearshore fish assemblages.

26. It has been hypothesized that a regional perspective may also be useful when monitoring the progress of a restoration project. Vegetation changes might occur that could be associated with regional influences such as rainfall or drought; or they could be caused by a localized phenomenon such as insect infestation. Faunal changes could be subject to similar multiple interpretations. Comparisons on a regional scale may or may not be relevant, depending on the circumstances. However, even if the driving force is widespread, the effect on individual wetlands may be very different due to dissimilar hydrological, soil, watershed, and upstream impoundment characteristics.

PUBLIC PERCEPTION AND THE REGIONAL PERSPECTIVE AS AN EDUCATIONAL TOOL

27. The involvement of the general public is important in the planning process of wetland restoration. In southern California, the role of the public previously had been minimized in the early planning stages of wetland restoration projects. They were brought in after design decisions had been made and design alternatives selected or reduced. In response to the sensitivity that the public recently exhibited to this matter, interested parties formed a public working group that took part in the technical design process for San Dieguito Lagoon. Members of the working group

SAN DIEGUITO LAGOON WETLAND RESTORATION PROJECT

EVALUATION CRITERIA AND DEFINITIONS[1]

EVALUATION CRITERIA: HABITAT DIVERSITY

Improve, preserve and create a variety of habitats to increase and maintain wildlife and ensure protection of endangered species.

(157) Restore a variety of habitat types including but not limited to:

- (67) saltmarsh specifically mudflats and vegetated intertidal
- (64) open water
- (46) brackish marsh
- (42) riparian
- (38) grasslands
- (31) upland buffer
- (23) coastal sage

(59) Optimize subtidal and intertidal areas in relation to biological and hydrologic goals.

(20) Increase aggregate acreage of wetland in the Southern California Bight.

EVALUATION CRITERIA: WILDLIFE CORRIDORS AND BUFFERS

Maintain the natural, open space character of the river valley with appropriate topography to support the ecosystem and viewshed.

Project should not contribute to a net loss of beach and sand north or south of the river mouth.

(57) Establish wildlife corridors.

(44) Provide average buffer of 300 feet, no less than 100 feet.

(38) Create buzzer zones along Del Mar Fairgrounds and Racetrack lands.

(32) Create wildlife corridors connecting the project with Gonzales Canyon, Crest Canyon, and the river valley east of the project.

(23) Create buffer zones.

---

[1] Numbers reflect scores given to each objective by the twelve members of the public Working Group.

Fig. 8. Example of decision criteria ranking by public participation group

ECOLOGY AND WATER QUALITY

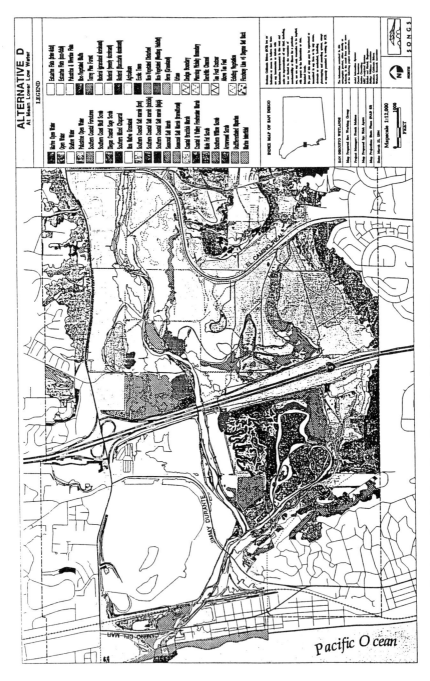

Fig. 9. An example of a wetland design alternative for San Dieguito Lagoon

began the process with specific agendas, which could have conflicted internally or with known wetland processes. A thorough review of the region's wetlands served to educate the group in wetland hydrology and ecology and provided lay-people with realistic expectations about what can or cannot be accomplished. With this knowledge, the group was able to rank its preferences in habitat, target species, species diversity, hydrological characteristics, and corridors and buffers (Fig. 8). Conflicts were more easily reconciled as a result of this regional background knowledge. This public participation has led to a set of alternatives (Fig. 9) that are presently being submitted as part of the Environmental Impact Review/Statement process that must be conducted for California and U.S. Environmental Protection Agencies. The scientific rationale for each alternative is understood by the proponent, the regulatory agencies, and the public through use of the regional background information. Even though each group and individual members of a group may support different alternatives, understanding, respect, and tolerance for the others has been achieved.

CONCLUSIONS

28. Thus, consideration of the regional perspective provides a valuable tool to coastal wetland restoration design. Specifically, the regional analyses provided the following recommendations for the San Dieguito Lagoon restoration:

• The lagoon entrance should be maintained to be permanently open to enhance diversity, productivity, and California Least Tern foraging areas and to retain the pelagic connection.

• Mudflat areas should be increased to provide greater attraction for migratory birds.

• The seasonal marsh area upstream in the southeast corner of the restoration area should retained. This area is still in a near-natural state and provides good bird habitat. Seasonal salt marsh habitat should be increased if possible because it provides good feeding and resting areas for migrating birds and refuge for other fauna during inclement conditions.

• California Least Tern nesting and foraging areas should be enhanced through the development of isolated, protected areas within the wetland and the provision of shallow water habitat.

• Pickleweed areas should be retained and expanded to further strengthen the Belding's Savannah Sparrow population and to move the wetland toward the reflection of historical conditions.

• Cordgrass planting experiments should continue, not in an attempt to provide enough area for clapper rail recruitment, but to provide a potential clapper rail stopover, to increase regional knowledge of cordgrass requirements, and to add another wetland to serve the limited gene pool.

REFERENCES

1. CHAMBERS GROUP, INC. Draft environmental impacts statement/environmental impact report (EIS/EIR) for the proposed Bolsa Chica Project. Prepared for the U. S. Army Corps of Engineers, 1992.

2. GIBBS, J. P. Importance of small wetlands for the persistence of local populations of wetland-associated animals. *Wetlands.*, 1993, 13 (1): 25-31.

3. JOSSELYN, M.N. and J. BUCHOLZ. Summary of past wetland restoration projects in California, p. 1-10. In M.N. Josselyn (Ed.), Wetland Restoration and Enhancement in California. California Sea Grant College Program. Report No. T-CSGCP-007. La Jolla, California, 1982.

4. KRAMER, S. H. Habitat specificity and ontogenetic movements of juveniles California halibut, Paralichthys californicus, and other flatfishes in shallow waters of southern California. National Marine Fisheries Service. Southwest Fisheries Science Center, 157pp., 1990.

5. MEC ANALYTICAL SYSTEMS, INC. San Dieguito Lagoon Restoration Project Biological Baseline Study, March-August 1992. Draft technical memorandum, San Onofre Marine Mitigation program. Prepared for Southern California Edison, 1993.

6. PERL (Pacific Estuarine Research Laboratory). A manual for assessing restored and natural coastal wetlands with examples from southern California. California Sea Grant Report No. T-CSGCP-021. La Jolla, California, 1990.

7. SORENSEN, J. Towards an overall strategy in designing wetland restorations. In M. Josselyn (Ed.), Wetland Restoration and Enhancement in California. California Sea Grant Program Report No. T-CSGCP-007. Tiburon Center for Environmental Studies, Tiburon, California, p. 85-95, 1982.

8. ZEDLER, J.B. The Ecology of Southern California Coastal Salt Marshes: a Community Profile. U.S. Fish and Wildlife Service. FWS/OBS-81/54. 100 pp., 1982

9. ZEMBEL, R., K. J. KRAMER, R. J. BRANSFIELD, and N. GILBERT. A Survey of Belding's Savannah Sparrows in California. *American Birds.* 42(5):1233-1236, 1988.

Water treatment systems – the need for flexibility

R. I. COLLINSON, Head, Environmental Assessment Division, WS Atkins Environment, A. G. HOOPER, Head, Environmental Hydraulics Division, WS Atkins Water, and M. HANNAM, Chief Environmental Scientist, South West Water Services Ltd

SYNOPSIS. The Water industry faces considerable challenges in meeting the regulatory requirements of the European Commission through the UK Government and its regulatory authorities, whilst providing investment returns for shareholders and responding as appropriate to the environmental aspirations of environmental groups and the public at large. The paper illustrate the way in which South West Water Services Ltd. has attempted to provide a flexible approach to a number of wastewater treatment systems. South West Water Services Ltd. have a stewardship role to enhance the environment which is reflected in the Companies Environmental Policy Statement and the commitment to achieve the Best Practical Environmental Option, having regard to feasability, affordability and environmental impact considerations. It has to be borne in mind that a considerable variation exists in the population served and that these numbers can be significantly increased by the seasonal influx of tourists; the resident population of approximately 1.5 million increasing to approximately 6.5 million during the summer months.

INTRODUCTION

1. The importance of coastal and esturial wetlands has been extensively described in other papers and is generally accepted by environmental groups, regulators and legislators. As a resource which is coming under increasing pressure from development, the protection of such areas becomes more urgently focused by conservation groups, especially as developments such as barrages, tidal power generation and land reclamation schemes threaten the diminishing pool of natural resources.

2. In addition to development issues, other pressures on wetlands arise from several sources, including recreational conflicts, flood defence measures, agricultural run-off, coastal protection measures and potential sea level changes and managed environmental impact. The fact that the habitats represented in these areas are subject to relative extremes of ecological tolerance, only compounds the combined effects of the various stress factors, which change in both relative and absolute terms.

3. Unfortunately, the protection of these habitats cannot be addressed in isolation since many of the factors causing stress have an historical basis and cannot be expunged or ameliorated overnight. In many cases, the use of the resource is considered legitimate, partly because there is no viable alternative and also because the dynamics of the wetland system can withstand managed exploitation. The crunch comes when it has to be decided what the tolerance levels of the environment are, within the context of sustainable development.

4. In an attempt to try to come to grips with the issue, the "conservation lobby", lead by English Nature (EN) and The Royal Society for the Protection of Birds (RSPB) have developed initiatives to investigate the significance of the coastal and estuarial resources of the UK. In October 1992, English Nature launched its *Campaign for a Living Coast* to highlight the potential threats to the over-exploitation of the coastal wetland areas. Part of their Campaign is represented by the *Estuaries Initiative*, an objective of which is to develop ways of managing estuaries in a sustainable way.

5. At the same time, the Water industry, through the actions of both the National Rivers Authority (NRA) and the privatised Water company plc's, are obliged to make provisions for the protection of the environment, both through the legislative procedures of the Water Act and to enhance the environmental management of the resources that are within their control.

6. As an element of this responsibility and following the recent interest in the development of coastal and estuarial resources under the English Nature coastal Campaign, the concept of integrated management planning has lead to a more coherent approach to the topic. We have seen a coming together of the different interest groups, often lead through the auspices of English Nature, but effectively supported by the Water industry and the relevant planning authorities.

7. Only a limited number of these estuary management plans (EMPs) have so far been undertaken, but they have served to prove how complex estuarial issues are and highlight the pressures which are exerted on wetland resources in this country, from a wide range of vested interest groups. Maintenance of water quality is paramount to the success of initiatives and proposals, and as such it is totally appropriate that the Water industry plays its role in the effective management of our wetlands.

THE WATER INDUSTRY PERSPECTIVE

8. The need to maintain and improve the water quality of receiving waters poses a number of issues which the individual Water companies are obliged to resolve. The challenge facing the Water industry is now the familiar dictatory between the environment and economics. The Water plc's have both legal obligations to meet and a group of shareholders to serve. Conservation groups and the general public also require that the duty to protect water quality - and hence wetlands - for the good of all, is met. The NRA control the implementation of the legislation, covering all aspects of the functions and operations of the plc's. However, it is down to the companies to determine, in conjunction with the respective regional NRAs, how they are best to meet their objectives, to the satisfaction of all parties.

9. In the first instance, it should be remembered that the companies are carefully monitored by the Director General of OFWAT to ensure that they operate in an efficient and cost effective manner, which means meeting legislative requirements at an affordable cost to the customer. In other words, the Water industry does not have a free hand to design improvements into wastewater schemes which satisfy all issues in equal measure.

10. Second, a further factor, acting effectively as a constraint, is the regulation enacted by the NRA over the majority of the companies functions. The regulation is based on a number of EC Directives, and UK Government interpretation of them, which gives a clear requirements for, amongst other activities, wastewater discharges. The problem is that, to some extent, the requirements of the EU are a moving target. Future legislation is anticipated to be driven by use-related criteria and as such is likely to impose further investment obligations on the Water industry. Consequently, installed systems must be sufficiently flexible to be

capable of appropriate modification, when legislative edicts dictate and financial constraints permit.

11. Associated with these responsibilities and controls, is the additional requirement of involvement with managing the coastal wetland resources that are often associated with the areas under Water company control. Whilst the disposal of appropriately treated waste water into receiving water bodies is recognised as a 'legitimate use', it makes little sense to endanger such fragile habitats by over-exploitation or lack of concern. Eventually, it is possible that any imbalance produced in the ecosystem will become attributable or regarded as a failure on the part of the Water companies. Consequently, it is in their interest to ensure protection of the natural resources. It is also appropriate to make use of the dynamics of ecosystems and habitat types, to assist in water purification processes and consequently to benefit as wide a range of interests as possible. Selecting appropriate treatment processes is therefore of paramount importance when considering the requirements for new scheme proposals. Consideration must be given to the nature of the receiving water, so that the potential implications of discharges on the local and regional natural environment are fully evaluated.

12. Before this can be accomplished, it is necessary to investigate the respective environments as fully as possible, to attain a degree of knowledge and understanding of the respective areas which could potentially be employed - or alternatively adversely affected -if we get the sums wrong. The investigations must include:

(a) collation and evaluation of water quality data;
(b) assessment of the dispersion characteristics of effluent;
(c) investigation of the flora and fauna of the effluent discharge area, together with consideration of the significance of the wider habitat context;
(d) review of the significance of fisheries and other commercial activities in the vicinity.

13. From the data and information collected, it is then possible to select a system of water treatment which is best suited to the overall requirements. Obviously, as discussed, the financial and engineering considerations are important in the selection procedure, but the significance of environmental issues cannot and are not ignored.

14. By way of illustration, we should like to briefly consider two specific engineering developments currently being undertaken by South West Water Services Ltd., in conjunction with WS Atkins. They form part of South West Water's major coastal water treatment programme of environmental improvement and demonstrate a range of engineering solutions, selected on the basis of legislative, environmental and economic criteria.

THE TREATMENT SCHEMES

15. The schemes being considered all involve, to varying degrees, the elimination of many small outfalls, the pumping of sewerage to treatment facilities and the discharge of treated effluent into the marine or estuarine environment. They form part of the UK Government undertaking to the improvement of bathing beaches, resulting in a South West Water Services Ltd. commitment of over £900 million in their 'Clean Sweep' improvement programme, designed to achieve European standards and to meet NRA objectives. All the effluent discharges are into or adjacent to areas which have intrinsic environmental significance, from an

ecological and amenity viewpoint. Appropriate levels of treatment and maintenance of plant performance are therefore of paramount importance for the successful sustained operation of the facilities.

16. The schemes are in different stages of development or operation and the scale of the facilities also varies considerably. The schemes are further influenced by the fact that of all the Water companies, South West Water Services Ltd. is perhaps faced with the most significant annual variation of treatment requirements. The region possesses almost one third of all designated and identified beaches in the country and as such, the significant influx of holidaymakers creates major fluctuations in treatment requirements. Consequently, it is essential that there is flexibility in the operation of the facilities, to allow for the considerable variation in baseload conditions. The schemes in question are:

(a) Taw Torridge Scheme (Bideford, Barnstaple, Northam, Cornborough, Westward Ho!) divided into northern and southern facilities;
(b) Thurlestone Scheme (Thurlestone).

Taw Torridge Scheme

17. Background The disposal of sewage from the towns of Bideford and Barnstaple, both several miles up the Taw Torridge estuary complex, has been the subject of argument and debate for a long time. A succession of schemes were proposed by the then South West Water Authority, none of which were implemented for a variety of reasons. Following privatisation, South West Water Services Ltd. initiated a programme of work and consultation to identify a scheme which would satisfactorily resolve this outstanding problem. A number of alternatives were identified, which ultimately resulted in the selection of a scenario addressing separate solutions for the treatment requirements north and south of the estuary complex (Ref No 1).

18. In addition, the Taw Torridge estuary is one of the first estuary systems that has been the subject of an EMP (Ref No 2), under English Nature's *Estuaries Initiative*. It is complex from a managerial viewpoint and physically diverse. The area is of outstanding ecological value, the coastline and intertidal parts of the estuary being designated Sites of Scientific Interest (SSSI), with the adjacent Braunton Burrows being a National Nature Reserve and a UNESCO Biosphere Reserve. The estuary is of particular importance for migrant and wintering waders and wildfowl, and is a candidate site for designation as a Special Protection Area under EC legislation.

19. Saltmarsh, shingle, sand dune, wet grassland and brackish marsh provide a variety of valuable habitats for nature conservation. There are also extensive areas of intertidal mud and sand flats especially along the Taw. However, much of the saltmarsh of the estuary has been reclaimed for agricultural use, resulting in accompanying drainage schemes and modifications to natural grassland habitats.

20. The area is also of considerable landscape value, a large part of the estuary lying within both an Area of Outstanding Natural Beauty (AONB) and a Coastal Protection Area. Much of the north part of the area has recently been designated a Heritage Coast. Both Barnstaple and Bideford are important historic centres and the area as a whole has a rich heritage. Braunton Great Field, for example, provides a rare example of an open field system which still retains its medieval character through the continuing practice of strip cultivation and drainage practices.

21. In addition to the physical characteristics of the area, the estuary provides an important recreational resource for the local population, sports clubs, activity

centres and holiday makers. A wide variety of activities are carried out including active water sports, for which acceptable water quality is a prerequisite These include water-skiing, windsurfing, surfing, sailing. rowing and canoeing. The area is also of key importance for informal recreation, particularly bathing and beach based activities.

22. The Taw Torridge estuary achieved a grade A classification both in the 1985 and 1990 national surveys of water quality, representing an improvement over the results attained during the equivalent 1975 survey. However, the estuary is still subject to a considerable number of outfalls, many of which are now the responsibility of South West Water Services Ltd. The result, is that whilst the majority of the beaches achieve quality criteria sufficient to meet the EC Bathing Beach Directive, Instow Beach in particular frequently fails the required standards. As indicated above, the need for improved facilities has been recognised for a considerable time, the present proposals having been developed only after considerable investigation from an engineering and environmental viewpoint and following considerable public debate on the topic. It is against this background, including consideration for the considerable environmental value of the area, that the potential schemes have been proposed.

23. The overall proposal for the area has ultimately been based on a two-system approach, with separate schemes being recommended for the areas north and south of the River Taw respectively (See Fig. 1).

24. The northern option. This consists of the following proposals:

(a) closure of Velator works, which currently discharges in the Braunton Pill;
(b) transfer of sewage flows from Braunton, via a new pumping station, to the existing sewage treatment works;
(c) modernisation and extension of those treatment works to meet the higher standards required by the NRA. This would include:
- preliminary screening treatment;
- primary settlement treatment;
- secondary biological treatment;
- consideration of tertiary treatment, dependent upon the results of future plant performance;
(d) discharge of treated effluent to the main tidal channel of the River Taw estuary through a new underground pipeline and diffuser system.

25. The above proposals, investigated through the initial environmental investigations described in para 12 and supported by an environmental impact assessment (EIA), have been accepted following a recent planning application and are in the process of being implemented. The scheme should result in the significant improvement of water quality downstream of the discharge, together with improvement in the air quality around the existing facility, complaints previously arising due to odours associated with sludge generation.

26. The southern options. These are still being considered, the potential schemes having been the subject of extensive debate, including public exhibition and consultation.

27. Under Option 1 (The Northam Option), a new sewage treatment works would be located between Long Lane and the Northam Burrows, the latter being designated a SSSI. The new works would treat flows from the existing fine screening works at Bideford and Westward Ho! The flow from Appledore, currently discharged into the estuary, would be pumped to the Westward Ho! fine screening works. Further treatment at Northam would then involve settlement to

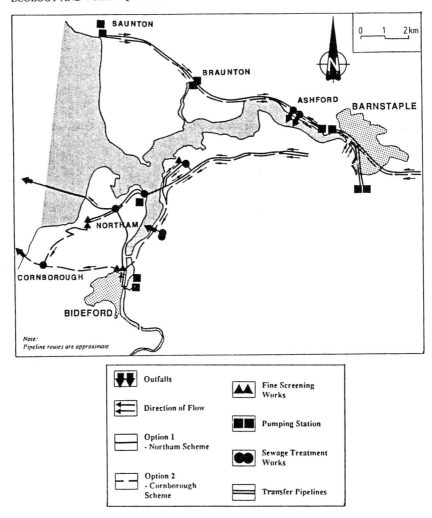

Fig. 1. South West Water Services Ltd. Option Proposals for Taw Torridge Sewage Treatment

remove solid material, before the treated effluent is discharged through a 4.5km pipeline laid below the seabed, into Bideford Bay.

28. Under this option, several additional improvements to local treatment works at Yelland and Westleigh would also occur.

29. Under option 2 (the Cornborough option), sewage flows would be taken from Bideford, Northam, Westward Ho! and Appledore areas, together with those from Yelland, and transferred to Cornborough for treatment. The treatment would include preliminary, primary, secondary and ultra-violet disinfection processes. Treated effluent would then be discharged through a 0.5km pipeline into Bideford Bay.

30. Under both schemes, the same environmental standards will be met, although there is considerable difference in the engineering approaches to be adopted. In addition to the considerable improvements in water quality, the chosen scheme will also have the flexibility to allow additional levels of treatment to be added to comply with any higher future environmental standards. This factor is significant, since tightening up of consent conditions is an inevitability as a result of the Urban Wastewater Directive.

31. Environmental considerations The northern proposal will discharge from Ashford sewage treatment works (STW) into the River Taw, which forms part of the Taw Torridge Estuary SSSI. Standards have been established by the NRA South West Region which will ensure that the effluent emissions from the STW will not have an adverse effect on the valuable ecological resources of the estuary. This not only includes the intertidal area, but also the adjacent saltmarsh and downstream locations that are significant for migratory birdlife. In addition, there will be considerable benefits for the recreational users of the estuary complex.

32. The southern proposal alternatives, whilst meeting the required effluent discharge standards, could have different influences on the natural environment. The Northam option is adjacent to the SSSI of Northam Burrows and would require the emplacement of the sea outfall through the designated area. This would likely be effected through trenching with concomitant restoration after pipeline installation. The line would probably be bored under the geologically significant Pebble Ridge, emerging some distance offshore. In total, a 4.5km pipeline would be created, effluent being discharged through a diffuser system into Bideford Bay. The dye tracking studies undertaken at the beginning of the study indicate that complete dispersion can be anticipated.

33. The Cornborough option, whilst not affecting the SSSI, would discharge through a considerably shorter outfall and as such would require a greater degree of treatment in order to meet NRA discharge consent conditions. It will also involve the provision of a pipeline through the coastal zone, a process which could involve trenching or bored tunnel construction techniques. In spite of the shorter outfall, the greater degree of on-land treatment will ensure that dispersion of effluent will be entirely satisfactory and will not cause any adverse amenity issues for the adjacent recreational bathing beaches at Westward Ho!

34. Other environmental issues such as area and type of land take, potential disruption to commercial and amenity undertakings, visual significance and intrusion due to construction traffic, are all being considered as part of the option selection process. Final selection will require a careful evaluation of the factors described, several of which vary considerably between the alternative scheme proposals. Employment of an EIA to support the proposals can also be used in the selection procedure.

35. To facilitate the selection, there is need for considerable flexibility in the engineering proposals available to the South West Water Services Ltd. design team. Short, medium and long term financial implications also have to be considered, together with the potential overall implications for the local environment. The views of the local authorities and residents also feature significantly in the selection procedure which South West Water Services Ltd. are presently undertaking. Improvements to the quality of the discharges entering the marine environment is an obvious prerequisite of the scheme. Minimisation of environmental disturbance is, however, also high on the list of selection criteria which are being applied to the exercise.

ECOLOGY AND WATER QUALITY

Thurlestone Scheme

36. Background Thurlestone is a small coastal village in the South Hams region of Devon, that is a popular summer tourist centre. Until the completion of the scheme described here, Thurlestone was served by an outfall that carried the village's crude sewage flow into the sea, just above the mean low water spring mark. It was typical of more than 100 other major outfalls in the south-west region that South West Water Services Ltd. is seeking to improve as part of the 'Clean Sweep' programme referred to above. The expansion and upgrading of the treatment works has meant that by the year 2001, a three and a half fold increase over original discharge quantities can be handled, thereby permitting appropriate treatment of the additional discharges due to the transient holiday trade inputs.

37. The outfall at Thurlestone only carried the flow from about 800 people, but it caused considerable local concern, due to the discharge being into the designated bathing water. There was also a combined storm overflow (CSO) that prematurely discharged to the small stream immediately upstream of Leas Foot Beach. Many complaints were made and although these were not new problems (as the outfall was first constructed in the 1930's), the number of incidents increased with the publication of unsatisfactory coliform concentrations, after regular monitoring was instigated in 1987.

38. Although the number of complaints continued to rise, and South West Water Services Ltd. had submitted its investment programme to the Department of Environment in 1987, Thurlestone was not scheduled for completion until 1993. However, it was recognised that it was in everyone's interest to try to solve the problem as soon as possible. Consequently, scheme options were considered at the end of the 1980's, including long sea outfalls against full biological treatment, and possibly disinfection techniques, with full evaluation on technical and financial grounds. In addition, however, the concerns of local customers and environmental pressure groups was also taken fully into account.

39. In considering the alternative schemes, a major concern for South West Water Services Ltd. and the NRA, was the fact that there was no significant watercourse outfalling to the coast that could offer reasonably satisfactory dilution of effluent. A very small stream outfalls onto Leas Foot Beach, whilst the larger watercourse passing adjacent to the treatment works, forms part of South Milton Ley (See Fig. 2). This waterbody is one of the best examples of a freshwater reedbed remaining in Devon. Situated in a shallow coastal valley below South Milton village, the Ley is separated from the sea by means of a continually moving sand bar. Adjacent to the sand bar, the Ley is totally influenced by fresh water. The Ley is designated a SSSI, under section 28 of the Wildlife and Countryside Act 1981, on habitat criteria and for its diverse community of breeding birds. Most of the Ley is owned and managed by the Devon Birdwatching and Preservation Society (DBPS).

40. Scheme Details Thurlestone itself is situated some 7km to the north west of the town of Salcombe and a similar distance to the west of Kingsbridge. As improvements were planned for the treatment facilities for each of these towns, consideration was given to a regional scheme, with either a major sea outfall and/or a single wastewater treatment works. Various possible locations were identified.

41. Technical considerations in favour of a regional scheme were the likely cost benefit of a larger scale facility, together with a possibility to either use an existing treatment works or provide one long sea outfall. Set against these potential advantages was the likely long timescale of implementing such a scheme, plus significant pumping costs.

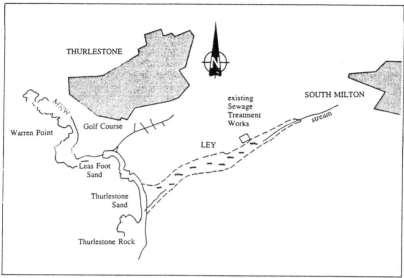

Fig. 2. Location of Thurlestone Sewage Treatment Works

42. It was decided that an interim measure (an attenuation tank) would be provided at the site of the existing CSO in Thurlestone, together with a fine screening facility. However, it was eventually decided that the regional scheme - and hence the need for the fine screening - would not be pursued Consequently, two local sites were considered for treatment:

(a) a greenfield site further up the Thurlestone valley; or
(b) an extension of the existing sewage treatment works (STW) adjacent to South Milton Ley.

43. It was decided that use should be made of the existing STW, though the NRA requirement for discharge consent was expected to be onerous. Consequently, a minimum of secondary treatment would be required, even if a sea outfall discharge was provided, due to the proximity of the bathing waters. Evaluation then showed that tertiary treatment of effluent to a standard suitable for discharge to the Ley would be less than the alternative cost of a pipeline and outfall.

44. Three solutions were shortlisted, namely:

(a) a conventional filter bed treatment;
(b) a system utilising rotating biological contractors (RBC); and
(c) a biological flooded aerated filter (BAFF).

Of the above, the decision was made to utilise a conventional filter bed system, based on noiseless operation, low maintenance requirement and favourable financial implications. The low noise level operation was of particular significance due to the proximity of the SSSI.

45. It was acknowledged that the conventional STW system would not attain the required standard of effluent, on its own. However, research indicated that tertiary treatment based on reedbed technology could meet the required enhanced

quality standards. Ultraviolet irradiation was considered, but this would only effect a reduction in the bacteriological levels, but not the organic and inorganic content of the effluent. Consequently, it was decided that a reedbed treatment system (RBTS) should be included in the scheme development.

46. Background to RBTS Given that RBTS is a relatively recently embraced treatment technique in the UK, it is worth giving consideration to its application in water treatment. Although RBTS and constructed wetlands have been used in the UK since the turn of the century, it was not until 1985 that significant interest in these processes arose. This followed a visit to Germany (Ref No 3) by a delegation from the Water Authorities and the Water Research Centre (WRc). The emphasis at that time was in the use of RBTS for secondary or even full treatment of sewage.

47. A number of the beds constructed for this purpose were commissioned using poor quality secondary effluent. The potential for using the process for tertiary treatment was soon recognised, but it was not considered cost effective to use soil media with the RBTS design parameters for secondary treatment. Work had been done by some of the Water Authorities and the WRc with gravel media (Ref No 4).

48. Most of the investigative work on RBTS has concentrated on the performance of suspended solids, BOD and NH_3 removal; phosphate removal has also been of interest. Very little work has been done on bacteriological reductions, although considerable benefit has been reported from one location (Ref No 5). Nevertheless, in spite of the extremely good treatment performances reported from the literature, it is still difficult to predict performance of RBTS's and a conservative approach to design should be used.

49. Of particular interest in the application of an RBTS at Thurlestone is the potential nature conservation benefit of such schemes. This issue is the subject of a further paper at this Conference (Ref No 6), but it is worth remembering that the application of RBTS by the Water industry is primarily for water treatment purposes. Generally speaking, in order to have an inherent conservation benefit, the area of reedbed required should be in the order of $1000m^2$ (Ref No 7). Even so, the creation of an RBTS is not primarily intended for habitat creation purposes. It is, nevertheless, fortunate that reedbed creation does offer ecological advantages.

50. The presence of the reedbed system at South Milton Ley could enhance the ecological value of the RBTS. Also, given that typically treatment areas range from $1-8m^2$ per person, the requirement for Thurlestone is sufficient to have additional nature conservation benefit. This situation will therefore be of interest to the management group for the Ley, since environmental groups such as the Royal Society for the Protection of Birds and Wildfowl and Wetlands Trust, have developed interests in wetland treatment systems. They are seen as reducing the impact of treatment plant on the natural environment and as providing additional wildlife habitat.

51. The Thurlestone STW The construction of the plant took place between April 1991 and August 1992, occurring in three phases involving:

(a) construction of the attenuation tank and pumping station at Thurlestone;
(b) supply and installation of the mechanical and electrical plant, including automatic desludging for the primary settlement tanks, grit and screenings removal and treatment, sludge storage, works pumps, washwater, controls etc;
(c) civils works for the STW.

52. During the first phase of the work, negotiations were under way with the NRA, with respect to the consent conditions for discharge to the Ley. This work was supported by various studies, including an ecological evaluation of the SSSI, the adjacent land-use and the potential management requirements to maintain the ecological and habitat diversity of the site. (Ref No 8). In addition, considerable effort was made to involve the local population and other interested parties, in the from of several public meetings and consultation processes. An outcome of these discussion involved the relocation of the new reedbed some distance from the original proposed position, in order to avoid an ancient hedgerow, a habitat used by dormice and the Cirl Bunting.

53. It was also felt important that the species of reeds to be planted (*Phragmites australis*) should not compete with the existing strain of the species present in the Ley. Consequently, with the permission of the DBPS, seeds were collected from the endemic population of reeds, to be cultured for planting once the preparation work for the constructed RBTS had been completed. Use was made of a gravel media reed bed, thereby reducing the area required relative to a soil based system and also ensuring subsurface flow. Iron rich material and limestone chippings were also included in the media, to maximise the potential for phosphate and nitrate reduction.

54. Full plant commissioning was completed on schedule in July 1992. Flows from Thurlestone were initially diverted whilst the initial seeding of the new filters took place over some weeks, utilising effluent from South Milton. As the filters became progressively established, so the amount of recirculation through the old works was reduced. Once commissioning was complete, a handover and training meeting was held with South West Water Services Ltd., Divisional operators and the works became operational in the truest sense.

55. Effluent monitoring by South West Water Services Ltd. and the NRA has begun in June 1992 and this has been continued to evaluate the efficacy of the new STW, including the performance of the RBTS.

56. STW Treatment Performance The reed bed was fed a poor quality final effluent from the old South Milton works for approximately seven weeks prior to commissioning the new biological filters to gain treatment benefit from the new reed bed, in order to afford protection to the Ley. Since that initial period, the performance of the works and particularly the reed bed has steadily improved; great consistency in performance has been attained during the last two years of operation.

57. The BOD, suspended solids and NH_3 concentrations within the final effluent discharges have all been extremely low, often similar to high quality river water. There is no doubt that the performance of the RBTS has been helped by the good quality effluent being received from the humus tanks and the biological filters, which only exemplifies the need to have completely effective plant operation in order to meet discharge requirements.

58. Results for the removal of nutrients have not been as encouraging. PO_4 and TON values have shown a mean removal rate of 55% and 26% respectively which, whilst a little disappointing, are in line with similar RBTS in other parts of the country.

59. The removal of micro-organisms, measured by indicator organisms of total coliforms, faecal coliforms and faecal streptococci, has been particularly interesting. At the early stages of the project when discussions were taking place with the NRA and interested groups, South West Water Services Ltd. was envisaging that the RBTS would remove one order of magnitude of microbes, in order to limit the pollutant loading to the Ley. From previous work carried out,

it had been shown that 1-2 orders of reduction could be achieved, but very little work had been undertaken on RBTS operating purely in a tertiary mode. Butler *et al* (Ref No 5) have reported three orders of reduction of total coliforms. However, the removal rates measure at Thurlestone consistently improved on these figures and sometimes have exceeded four orders of reduction.

60. An overall indication of performance of the RBTS are presented in Table 1, the range of effluent quality parameters being expressed (in the last column) as a comparison of mean effluent results against influent readings to the RBTS.

61. Environmental considerations The success of the scheme should not be judged solely against the effluent chemical and bacteriological quality performances alone, although it should be recognised that the scheme as designed has fully met the main objectives of providing consistent compliance with the EC Directive limit values for the indicator organisms in the bathing waters over the last two years. The constructed wetland has developed and provided an extension to the natural habitat of the Ley, both as a food source and as a visually integral part of the local landscape.

Table 1. Effluent Quality Parameters for Thurlestone STW operation

Parameter	Range	Mean	Mean Removal Rate (%)
Total BOD (ATU) (mg/l)	0.4 - 19.0	4.1	68
Soluble BOD (mg/l)	0.1 - 19.0	3.5	55
Suspended solids (mg/l)	0.5 - 13.0	3.4	87
Ammonia (mg/l)	0.5 -13.0	0.7	76
Nitrite (mg/l)	0.05 - 1.2	0.5	62
Total Oxidised Nitrogen (mg/l)	5 - 20.7	15.2	26
Phosphate (mg/l)	0.01 -8.0	5.6	55
Total coliforms (10^6/100ml)	0 - 0.54	0.025	99
Faecal coliforms (10^6/100ml)	0 - 0.24	0.012	96
Faecal streptococci (10^6/100ml)	0 - 0.033	0.0015	93

62. The existing Ley shows no sign of adverse impact, though it must be recognised that ecological change often takes time to manifest itself. Having said that, however, the fact that no detectable modification is evident is encouraging. Indeed, water quality monitoring of the stream upstream of the outfall has often indicated that the works discharge and the stream quality to be of the same order. The concentration of indicator organisms in the South Milton Stream downstream in the Ley has significantly reduced over the two years since the works have been in operation.

63. Of further benefit is the fact that low maintenance requirements have resulted in reduced visits to the site by South West Water Services Ltd. personnel. To date, the only maintenance on the bed has been preventing *Phragmites* growth over the inlet and outlet channels, and the footpath that passes across the centre of the bed to the sampling point.

CONCLUSIONS

64. A number of wastewater treatment systems have been referred to, in an attempt to illustrate the range of facilities that are available to the Water industry and to which individual Water companies must refer, in order to provide a flexible approach to the variety of communities which they must serve. Whilst the EC and Governmental regulation dictates the standard of effluent treatment which must be applied to meet environmental quality standards, it is the responsibility of the Water companies to meet these requirements in the most cost effective, and yet environmentally acceptable way possible.

65. Although the financial and engineering implications of the schemes are of considerable significance in selecting preferred options, environmental considerations have become equally important. Large scale facilities must integrate into the overall management plans of regionally important areas; local schemes must integrate into the endemic environment, so as not to cause undue disruption. Wherever possible, environmental enhancement should be the goal of all schemes, not just because of legal dictate, but also to meet the reasonable aspirations of local residents and general public alike. The above examples have, it is to be hoped, demonstrated that all the above objectives are possible. All that is needed is the initial appreciation and vision that 'environmentally friendly' schemes can have significant engineering benefits as well.

REFERENCES

1. South West Water Services Ltd. (1992) 'Clean Sweep' for the Taw-Torridge Estuary & Bideford Bay.
2. WS Atkins (1993) Taw Torridge Estuary Management Plan. A report to a consortium of local and regional authorities, conservation organisations and the Water industry.
3. Boon, AG (1986) Report of a visit by Members and staff of WRc to Germany (FGR) to investigate the root zone method of treatment of wastewaters. WRc Report 376-s/l.
4. Fox, I and Wharfe, R (1989) Recent advances in sewage purification - gravel reed bed treatment systems. Paper presented to the Institution of Water and Environmental Management, South East Branch Symposium: New Service and Technology - Developments for the Future Water Industry, Gatwick, 20 April 1989.
5. Butler, E., Loveridge, RF., Ford, MG., Bone DA and Ashworth, RF (1990) Gravel bed hydroponic systems used for secondary and tertiary treatment of sewage effluent. J.IWEM, 1990, **4**, (3), 276-284.
6. Merritt, A (1994) The wildlife value and potential of wetlands on industrial land. Session 6, Wetland Management, International Conference, The Institution of Civil Engineers, London, UK, June 2-3 1994.
7. Coombes, C (1994) The use of reed-bed treatment systems for small communities and developments. Paper presented to Africa 1994 Water and Environment Conference, Accra, Ghana, April 1994.
8. South West Water Services Ltd. (1991) Thurlestone Sewage Disposal, South Milton Ley Environmental Baseline Survey. Prepared on behalf of South West Water Services Ltd. by WS Atkins Environment, November 1991.

DISCLAIMER

The views presented in the paper represent those of the individual authors concerned and do not necessarily reflect those of WS Atkins or South West Water Services Ltd., the latter addressing such issues in their published Company Environment Policy document.

The wildlife value and potential of wetlands on industrial land

A. MERRITT, Research Officer, The Wildfowl and Wetlands Trust

SYNOPSIS. The Wildfowl and Wetlands Trust undertook a project during 1992-94 looking at the wildlife value and potential of wetlands owned and managed by industry in the UK. This work indicated that there was great potential for increasing the wildlife value of such wetlands. This paper illustrates a few examples of how improvements could be achieved.

INTRODUCTION
1. The Wildfowl and Wetlands Trust has recently completed an eighteen month project, sponsored by British Coal Opencast, which looked at the actual and potential wildlife value of wetlands in the management or ownership of industry in the UK. Through its involvement in related projects the Trust had become aware that:
   a) wetland creation and enhancement schemes were becoming increasingly popular;
   b) an increasing interest in the use of constructed wetlands, especially for the treatment of wastewater and the storage of flood water, offered opportunities for encouraging wetland wildlife;
   c) major changes in land-use policy, releasing areas previously used for agricultural production could result in more land becoming available for habitat creation.
2. The project considered a wide range of wetlands with an industrial connection and potential as wildlife habitat, including water storage reservoirs, silt storage lagoons, constructed wetlands, reclamation schemes and examples of habitat creation on non-operational land. Drawing on the project's findings, this paper provides a few examples of how future 'industrial wetlands' could be enhanced for wildlife, using three of the most widespread categories of industrial wetland (reedbed treatment systems, flood storage wetlands and wetlands created as part of a reclamation scheme) to demonstrate the breadth of opportunity available.
3. Information gathered on numerous sites indicated that many such wetlands had not realised their full potential in terms of wildlife habitat, and that there were often further opportunities to encourage wildlife. Many projects designed to encourage wildlife would provide greater benefits if more thought had gone in to clarifying the wildlife objectives, ensuring that they were appropriate to both the location and the type of wetland involved. Fuller consideration of the whole range of factors that influence the ecology of a wetland would often have provided further benefits.

REEDBED TREATMENT SYSTEMS

4. During the past 15 years or so, a great deal of interest has been generated in the use of reedbeds for treating wastewater and effluents. Despite often being sold on their environmental benefits, remarkably few studies have been published on the wildlife value of reedbed treatment systems. In many respects, these systems mimic natural reedbeds which are valued as the habitat of various bird species. Most of the information on their actual usage by birds has been anecdotal, but does suggest that small beds can attract species such as reed warbler, sedge warbler, reed bunting and water rail.

5. In order to maximise the wildlife potential of reedbed treatment systems they should be designed to replicate the features of important natural reedbeds:
* Areas of open water within and adjacent to the reeds.
* A mixture of areas of dry and flooded reedbed.
* Complementary habitats, such as carr and fen, in close proximity.
* Large size - the larger the reedbed the greater the range of specialist birds and other wildlife that could potentially occur; while Reed Warblers will nest in reedy fringes, Bitterns mainly breed in reedbeds larger than 24 ha.

6. Most reedbed treatment systems constructed in the UK to date have been small (the largest is 18 ha, but most are less than 1 ha) and/or of an experimental nature. The minimum size for a reedbed system will primarily be determined by local topography and treatment requirements. Over-sizing treatment cells will tend to improve effluent quality as well as increasing the opportunities for including in-site variation such as ponds and willow scrub. The design of a system incorporating habitat features will need to ensure that it is not prone to short-circuiting or channelling of the through-flow.

7. Water features, as well as areas of wet reed, are most easily incorporated into overland flow systems and for this reason these systems are to be preferred to sub-surface flow systems from the perspective of wildlife habitat. The majority of sediments tend to settle out in the upper 15% of a treatment cell. Consequently, water features should be located away from the inflow and preferably about half way or more along a cell. The simplest open water feature that can be incorporated is a 2-3 m wide ditch which could provide feeding sites for reed warblers, water rails and even bitterns and bearded tits. Where space permits, wider expanses of water would be preferable as they should attract larger numbers and a wider selection of waterbirds. Such water features should be aligned at right angles to the flow to ensure that they do not encourage channelling. A band of emergent vegetation should always be left between the water feature and the outflow in order to trap any floating sediments, including algae, which could interfere with the outflow mechanism and affect readings related to the discharge consent. A simpler option for creating a reed-water interface, particularly where sub-surface flow is being employed, might be to pass the effluent from the reedbed cell into an adjacent reed-fringed lagoon.

8. The addition of water features need not lessen the efficiency of a wetland in terms of water treatment. Several reedbed-pond-reedbed systems operate in the USA and show enhanced ammonia removal over comparably sized pure reedbed systems. In Maine, USA, some treatment systems have been

constructed in the form of a sequence of wetlands; effluent passes from a sediment ditch through grass plot, reedbed, pond and, lastly, grass plot or marsh again, offering a potentially excellent habitat complex for wildlife. These wetlands are used for the control of nutrients and sediments in run-off from potato fields and achieve impressive levels of treatment including removal of 90-100% of both suspended solids and BOD (ref. 1).

FLOOD STORAGE WETLANDS
9. Ponds, lakes and other basins used to store flood waters can be incorporated into many situations and offer tremendous potential as wildlife habitat. Many flood storage wetlands are constructed in floodplains. In such situations there is little natural fall between the inflow and outflow encouraging the creation of extensive shallow basins which can make excellent wildlife wetlands. The value of such basins can be enhanced through the incorporation of a few minor features, such as deeper hollows to retain water during dry periods and mounds to act as islands.

10. A feature of many flood storage wetlands is a highly fluctuating water level which results in an unnaturally stressful environment for many forms of aquatic life. Such regimes often support impoverished plant and animal communities composed of a few more adaptable species such as Fennel-leaved Pondweed and chironomid (midge) larvae. However, such wetlands can be of value to a wide range of waterbirds. Willen Lake, probably the largest purpose-built stormwater balancing lake in the UK (68 ha), is one of a series of balancing lakes designed to take surface run-off from Milton Keynes. The North Basin of the lake has many features associated with a classic bird lake: large size, extensive shallows (mainly <1.5 m deep), eutrophic water (typical summer pH 7.8-8.5), good growths of aquatic plants and a large island (2.5 ha). Not surprisingly, therefore, it quickly established itself as an important site for waterbirds. It regularly attracts up to 2,500 wildfowl in the winter and supports an interesting selection of breeding birds, including 35-40 pairs of common tern.

11. Where only shallow fluctuations in water level occur it might be possible to encourage a wider range of wildlife, including dragonflies and other aquatic insects, amphibians and a rich variety of aquatic plants. This has been achieved at Potteric Carr Nature Reserve, South Yorkshire, where a flood storage facility has been superimposed upon an existing wetland complex including large areas of reed fen and open water. This system, which was put in place between 1978 and 1980 at a cost of over £1 million, utilises two major electrically operated pumps and offers sufficient capacity (230,000 m^3) to cater for a one in fifty year flood. The overall layout and operation of the system are strongly influenced by ecological objectives and exhibit several features of benefit to wildlife:
* The use of three large, discrete flood storage areas, which occupy compartments bounded by a series of railway embankments and cover twice the surface area (32 ha) required simply to fulfil the engineering requirements
* Water level fluctuations are limited to no more than 0.4 m (i.e. a depth of between 0.4 and 0.8 m) so as not to inhibit the growth of the reedbeds which provide one of the most important habitats.
* The use of three discrete storage areas means that one of the areas (Piper Marsh) is only affected by the most

severe storms, and, therefore, offers a particularly stable habitat.
* Only one of the storage areas (Balby Carr) receives the contaminated stormwater carried in a major drain that receives effluent from a sewage treatment works just upstream.
* The pumps can be used to top up the storage areas during periods of dry weather as well as for moving stormflows. They also afford flexibility to the scheme; the water levels could be increased if this was found to be more beneficial to the reed fen and other wetland habitats being conserved.
* A control chamber allows the control of the water passing under gravity from the primary storage area of Low Ellers to the secondary area of Piper Marsh. Under normal conditions the Yorkshire Wildlife Trust, who manage the reserve, seek to minimise the rate of through-flow in order to increase the purification of water entering Piper Marsh.

* The facility to flood the fields on the north and north-east sides of the reserve. (This facility has not been used to date.)

RECLAMATION SCHEMES AND WETLAND SUBSTRATES

11. A common short-coming of wetland creation schemes is a lack of consideration given to the physical characteristics of the new wetland, such as water depth, water level fluctuations and, particularly, substrate quality. The porosity, structure and chemical composition of soils and other substrates underlying a wetland significantly influence its character in a number of ways:
a) they determine permeability and sub-surface water movements;
b) they provide both a medium in which benthic (bottom-dwelling) invertebrates can live and a source of organic matter on which they can feed;
c) they provide a source of plant nutrients and a medium in which plants can root;
d) they influence the chemical nature of the overlying water.

12. While most wetland designers ensure that they include an impermeable liner where necessary, few consider the suitability of a substrate as a medium for plant growth and as habitat for benthic invertebrates.

13. While many habitats, including heathlands and flower-rich grasslands, are inhibited by nutrient-rich soils, good soil structure is an important component of wetlands designed for waders and other birds that feed on earthworms and benthic invertebrates. Good top-soil will also be beneficial in areas that are likely to receive heavy trampling pressure and hence require a robust sward. The choice of substrates available will vary according to a site's former use. At one extreme is opencast coal, the excavation of which can produce up to 25 times more waste than coal, exposing a range of different geological strata. Top-soil is often in limited supply on reclamation sites for a number of reasons: only a shallow depth across the original site; loss through mixing with other substrates; damage through compaction by vehicles; degradation resulting from storage over a long timescale; an increase in surface area of the reclaimed site compared with the original site. As well as seeking to maximise the resources available, consideration should be given to prioritising where to use the

top-soil across a reclaimed site. Taking into account the structure and nutrient status of the soil(s) available, an assessment can be made as to which of the proposed habitats it will benefit most.

14. The 12 ha North-west Lake at St Aidans, West Yorkshire, was created in 1987 as part of the first phase in the restoration of a huge opencast coal site. The shallow lake was intricately contoured with the intention of attracting a wide selection of waterbirds. In the first few years, waders were a prominent feature of the lake's bird life; 29 species were recorded. Many of the migrant waders stayed for only a few hours or even minutes, suggesting that although the lake was visually attractive to these species, it did not offer them adequate food. This undoubtedly reflects the predominance of a clay substrate lining the lake, a typical feature of many wetlands created through reclamation schemes. Unfortunately, only small quantities of top-soil were spread across parts of the lake bed.

15. In most situations it will be possible to overcome such short-comings. Where top-soil is in short supply consideration can be given to increasing the levels of organic matter. Sub-soils and lower strata contain little or no organic matter and hence offer only low levels of plant nutrients and little food for still water benthic invertebrates. A wide range of substances, including manure, compost, grass clippings, hay and straw, have been used to overcome such deficiencies, with varying degrees of success. Ideally, organic matter should be added to the floor of a wetland before it is flooded (or while it is kept drained). By giving such materials time to rot down aerobically, the flush of nutrients into the water should be reduced, and the material is less likely to float and/or wash away. Unlike many sources of organic matter, straw contains relatively low levels of plant nutrients and has the added advantages of releasing chemicals that inhibit the growth of algae and offering good structure as habitat for benthic invertebrates. However, straw must be soaked for several weeks, or even months, to prevent it floating, and therefore considerable care needs to be taken over its introduction. Another good method of introducing a textured organic layer is to sow a 'sacrificial crop' or green manure before flooding the wetland.

16. In eutrophic waters similar benefits may be obtained simply by increasing surface texture. Smooth substrates such as bedrock and compacted clay can be covered with a sandier or more soil-like material in order to provide a suitable medium for rooting plants and aquatic invertebrates. Coarser materials, such as stone or pebbles, can be valuable in waterbodies lined with mobile silts where they increase bed stability. In catering for invertebrates a covering deposit need not be very thick; most chironomid larvae, for example, live in the upper 100 mm of substrate, while a soil capping of 300 mm depth will be sufficient to cater for the roots of most aquatic plants. Simply breaking up compacted clays using a rotovator can increase densities of chironomid larvae and other benthic invertebrates.

17. At a coarser level, various materials, from bundles of brushwood to derelict gantries, can be placed on the bed of a waterbody to create an artificial reef. Such reefs can provide attachment points for invertebrates and algae and sheltered water and refuges for fishes and other free-swimming creatures.

18. These few brief examples illustrate how fairly minor

design considerations can influence the value of wetlands to wildlife. In the hope of encouraging wetland designers to consider these, and many other, suggestions in the future, The Wildfowl and Wetlands Trust will be publishing a manual on the enhancement of industrial wetlands for wildlife later this year (ref. 2). These suggestions draw heavily upon our current knowledge of wetland ecology. As our understanding of wetlands and their ecology increase so it should be possible to offer more refined guidance in the future.

REFERENCES
1. OLSON R.K. and MARSHALL K. The role of created and natural wetlands in controlling nonpoint source pollution. Proceedings of a US EPA Workshop. Ecological Engineering, 1992, vol. 1, 1-170.
2. MERRITT A. Wetlands, industry and wildlife. A manual of principles and practices. The Wildfowl and Wetlands Trust, Slimbridge, in press.

Management of the Essex saltmarshes for flood defence

D. J. LEGGETT, Coastal Processes Engineer, and M. DIXON, Senior Engineer, National Rivers Authority, Anglian Region

SYNOPSIS

The NRA maintains four hundred and forty kilometres of sea walls in Essex. Sixty per cent of these flood defences are fronted by saltmarsh, a natural wetland habitat that helps maintain the defences' integrity. As the saltings of Essex continue to erode the value of them as soft defences becomes increasingly apparent. The Essex Local Flood Defence Committee (LFDC) formally recognised the need to manage saltmarshes effectively in 1986, and implemented a programme of research and development that continues today. The Essex Saltmarsh Research Programme provides a practical example of integrated coastal management. It has provided new techniques for managing flood defence in Essex and developed policy towards saltmarshes.

INTRODUCTION

1. Management of the coast has occupied the energies of man for centuries and has included the reclamation of wetland areas for agricultural, industrial and urban use. Sea walls have protected the reclaimed land from flooding in a piece-meal way to meet people's short term needs and expectations. The problem identified by officers of the NRA's predecessor organisation was that the remaining Essex saltings were (and still are) eroding. It was apparent that, where saltmarsh became narrower, defences were subjected to greater forces than their design capability (ref. 1); highlighting the role of saltmarshes as an integral part of flood defences.

2. The saltmarshes of Essex provide a microcosm of national issues facing the coast; they are under threat from landward by development and seaward by erosion. The policy of protection of land is a reversible one, economics have changed reducing the perceived value of agricultural produce and consequently the ability to financially justify defence has become marginal, for some locations (ref. 2). This is not just a question of retreat, advance, or maintaining the status quo, the approach to management of the coast is at the heart of the debate. Progress towards integrated coastal management is constrained by the

variability in legislation and numerous organisations that have interests or responsibility on the coastline (ref. 3). An approach is required that recognises the restrictions of the coastal manager to a specific goal, and that conflict on key issues is not desirable. Conflict should be resolved at an early stage through increased awareness, understanding, and education. When emergency situations arise, and there is a threat to human life, democratic management is compressed; a principle accepted by all coastal managers.

3. To understand the role of saltmarshes in flood defence it is important to consider why they exist and how they have changed in the past. The recent history of saltmarsh use has moulded our perceptions towards coastal wetlands; they are valuable assets for flood defence and one of Britain's most threatened habitats (ref. 4). We should learn from history and try to avoid repeating the mistakes of the past. In order to achieve this it is necessary to gain greater understanding of how such natural systems function and how flood defences change them. This requires fundamental research and the application of that knowledge to develop new management approaches.

BACKGROUND

4. Saltmarshes in Essex are formed by muddy sediments being deposited under low energy conditions. Mudflats form between Mean Low Water (MLW) and Mean High Water Neaps (MHWN) (Fig.1).

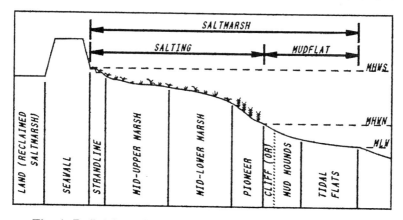

Fig. 1. Definition of saltmarsh zones on the Essex coastline.

Tidal flats form the most extensive part of mudflats, but at their upper limit they have either low cliffs or sloping mud-mounds. These two types of geomorphic feature can be created by either erosion or deposition. Vegetation can establish itself when mudflats accrete to a level around MHWN, and has an upper limit around Mean High Water Springs (MHWS). Vegetation is classified (ref. 4) into Pioneer, Low-mid, Mid-upper, and Strand line zones within this vertical range (Fig.1). Saltmarshes are very susceptible to changes in their environment and are therefore early indicators of natural, and man induced, impacts. The saltmarshes of Essex have witnessed many changes over history and this helps us to understand their distribution today.

Sea Level Change

5. At the end of the last major Ice Age (between 10,000 and 7,000 years BP) sea level rose and reworked sediments left behind as the ice melted, at the same time isostatic re-adjustment caused south eastern England to dip downwards further exacerbating relative sea level rise. In geological terms this represents only a medium scale change (ref. 5), with eustatic sea level rising only a matter of a few metres since 5,000 years BP (ref. 6). Sea level has, however, fluctuated in response to smaller climatic change. The 'Little Ice Age' from 1550 to 1700 caused sea levels to fall whilst there was a notably warm period from 1920 to 1950 (ref. 7). Sea level change has seen saltmarshes forming, migrating, drowning, and forming again further landward (ref. 8-10). The geological record shows there are macro processes effecting the Essex coast and that change is inevitable; as Pethick commented in 1984 (ref. 8) "...an increase in tidal range may be responsible for the observed saltmarsh erosion on the Essex coast...", a decade later the process of erosion continues.

Human use of saltmarshes and associated flood defences

6. Human activity on the Essex coast has a very long history. The archaeological record (ref. 11) shows Bronze Age (3,800-2500 years BP) settlements that are now beneath the Essex mudflats, and Roman (40 to 410 AD) inter-tidal fish-traps submerged at low tide, abandoned to the rising sea. It was, however, the medieval peoples (500-1500 AD) whom first recognised the potential of using saltmarshes for grazing sheep and cattle, hunting duck, and producing salt. Grazing became a widespread practice so that by 1086 the Essex saltmarshes sustained 18,000 sheep (ref. 12). Construction of causeways allowed people and livestock to move across the marshes and escape the flood tide. Over time the causeways were added to so that by 1200 they had developed to flood defences with drainage to landward (ref. 13). With improvements in technology in the Nineteenth Century ninety per cent of the salting area was reclaimed. The demand for food generated significant income for landowners, and allowed investment in further reclamation for arable crop production and the strengthening of defences to protect their assets from flooding (ref. 14).

7. The historical response to flooding has been to build defences ever higher; flooding in 1845, 1856, 1874, 1881, and 1897 all led to raising sections of defences. Further flooding in 1906 resulted in the appointment of the first Royal Commission into flood defence, in the United Kingdom, which reported in 1911 and put forward the proposition that either sea level was rising or land levels were sinking (ref. 15). In

1930 the Land Drainage Act apportioned responsibility for flood defences, and legalised precepts to finance them, but this did not prevent the 'Great Flood' of 1953 killing 119 people, 8,928 farm animals, making 21,000 families homeless, and causing the loss of 61,500 hectares of agricultural productivity in Essex. Following the Great Flood the Waverley committee was established which reported in 1954 (ref. 16) with recommendations including: development below the 5m contour should be precluded, the role of vegetation in flood defence should be investigated, sluices should be improved to evacuate flood waters faster, landowners should pay for the protection of their assets, and a single body should be responsible for planning in the coastal zone. Not all the recommendations were supported but extensive refurbishment, and re-design, of flood defences was undertaken and these are still being maintained and improved today.

8. In order to make progress it is necessary to understand the inter-relationship of the environment and the defences; in this way it may be possible to develop sustainable management for the coast. The Essex Saltmarsh Research programme (ref. 17) has been instrumental in allowing such progress to be made for managing flood defence in Essex.

THE ESSEX SALTMARSH RESEARCH PROGRAMME
Instigation

9. In 1985 the responsible flood defence officers undertook an extensive literature review of scientific papers, reports, and other archive material relating to saltmarshes and their management. Armed with this knowledge it was possible to propose areas for the initial research. The proposals came from a group traditionally perceived to have conservative views, the Engineers, and were supported by the Essex LFDC. The initiation of the programme through a well established institution (with an understandably cautious approach to new departures) is to their credit. The LFDC contributed £150,000 to research work and wholly funded practical experiments, and the European Community provided £250,000 for research into related conservation and environmental issues. The initiation of this programme shows us the farsightedness of individuals and the time lag between identifying an approach and reaping the rewards.

10. Definition of a unified programme of research enabled researchers to focus upon their particular discipline whilst placing responsibility for guidance, integration, and implementation with the coastal managers and LFDC. The research and development programme utilised specialists from differing disciplines to provide information on saltmarsh processes, vegetation, pollution, marine invertebrates, and birds. The research programme continues today and is fundamental to improving practical knowledge for management of flood defence and flood control.

MANAGEMENT SOLUTIONS

Saltmarsh Processes Studies

11. Changes in sea level have altered saltmarshes over long time scales but in 1986 there was little understanding of the impact of contemporary physical processes. Research by the University of Hull since 1986 has considerably extended the understanding of how saltmarsh processes work.

12. The inter-relationship of vegetated saltmarshes and mudflats are extremely complex (ref. 18). In Essex saltings continue to grow vertically, at a rate just sufficient to keep pace with present relative sea level rise, but also respond by eroding landward at their leading edge (Fig.2).

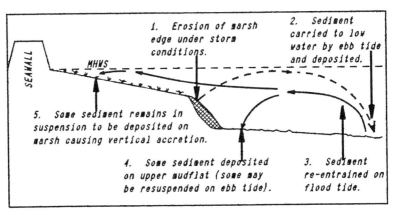

Fig. 2 Sediment movements across the saltmarsh profile.

Exposure of a site to the open sea leads to the vegetated surface eroding under storm conditions and the removed material being deposited on the mudflat. Under calmer conditions sediment is removed from the mudflat and returned onto the marsh surface (Fig.2). The research findings demonstrate that natural forces are capable of modifying the profile in a similar way to sandy beaches; the greater the energy the flatter the profile (ref. 19). This response presents the most effective long term solution for flood defence; a natural system that can control the impact of flooding.

13. Waves are the main cause of vegetation removal at the marsh edge. Waves cause vegetation to gyrate around their roots, causing scour at the plant stem leading to root exposure and ultimately removal. The remaining hole erodes further by wave action and the marsh edge decays. In more sheltered locations tidal energy also plays an important role. Sea level rise causes an increase in tidal energy and estuary and creek systems respond by infilling their sub-tidal channels and increasing their channel width. The truncation of saltmarsh creeks by reclamation increases tidal scour locally, widening the remaining creek and cutting a

new channel along the line of defence. Reclamation and relative sea level rise have prevented the system attaining dynamic equilibrium; creating a trend of increasing erosion.

14. The improved understanding of coastal processes has allowed recognition of the physical problems faced, and wider consideration of the approaches available to best manage the environment. Extensive management is required and considerable expense incurred where it is not possible to allow natural processes to prevail. Where it is necessary to intervene coastal managers must try to work with the natural system rather than against it.

Vegetation Studies

15. Vegetation helps in trapping and strengthening sediment, if vegetation dies back (rather than being eroded away) the natural system is unstable. Research by the Institute of Terrestrial Ecology (ITE) concurred with the distribution of vegetation loss identified through the process studies. Most importantly it identified a landward shift in the whole vegetation profile so that at a given point in space vegetation is changing to 'younger' (and hence lower) marsh. This is an indicator that the saltmarshes of Essex cannot surmount the effects of relative sea level rise and that, as time passes, they are providing less protection to flood defences.

16. The research also identified more localised losses of vegetation by the erosion of redundant oyster pits, smothering by mats of algal blooms, and limited damage by grazing sheep and Brent geese (ref. 17). The ITE also considered the effects of Tributyl tin (TBT) which was used widely as an anti-fouling agent for boats. High concentrations of TBT can kill, or prevent the establishment of, some pioneer plant species. High level concentrations in some localised sediments could account for the lack of vegetation in those areas.

Pollution Studies.

17. The concern raised by TBT led to the investigation of other pollutants including heavy metals, herbicides, insecticides, and Polychlorinated biphenols (PCB) by Imperial College, London (ref. 17). This work is on-going but has identified that under heavy rainfall events 'shock' loadings of pollutants are flushed from terrestrial sources through drainage sluices and onto the saltmarsh causing the possible demise of vegetation and hampering re-colonisation. It is also hypothesised here that these chemical pollutants, in association with fresh water and detergents, may cause de-flocculation of marsh sediments enabling them to be readily transported away. The effect of this pollution is to leave a low basin, with no vegetation, directly in front of some embankments allowing increased wave action and greater tidal energy to propagate to the flood defence.

Invertebrate and Bird Studies

18. Invertebrate studies by the University of Essex (ref. 17)

identified 23 taxa (species) in the mudflats and bottoms of creeks, and 103 species on vegetated saltmarsh. The salting species are instrumental in breaking down dead vegetation and turning it into carbon and nutrients that provide food for invertebrates in mudflats and creeks. Although the diversity of species in the mudflats is low the biomass is high, for example counts of 15,300 Hydrobia ulva (mud snails) per square metre were recorded. Invertebrates provide an important function in stabilising deposited sediment as they digest materials and glue them together with mucus (ref. 20). During calm periods and in summer (when numbers are highest) this process aids vertical accretion of mudflats. This sediment is not capable of withstanding wave attack but helps to dissipate wave energy and provides sacrificial material that is eroded by winter storms.

19.   Molluscs also assist in sedimentary processes, not only do they siphon large volumes of water (gluing sediment together) but also contribute their shells. Pollutants have reduced mollusc numbers and size accounting for the low contemporary input of shell material into the coastal system. The remaining vestiges of shell ridges in Essex are mainly the product of erosion of relict deposits that are driven into ridges under storm conditions. The ridges either protect the front edge of the saltings or ride up onto the vegetated marsh protecting it further landward (ref. 21).

20.   The Royal Society for the Protection of Birds (RSPB) carried out a bird survey of the Essex coast under the research programme. The RSPB reinforced the international importance of the Essex saltmarshes for migratory birds and uncommon species. Birds are not only an important part of the environment but also an economic asset for some landowners through bird watching and wild fowling. Birds feed on invertebrates and algae on mudflats and use the saltings as nesting and roosting sites. This demonstrates that saltings provide an important safe haven for invertebrates to breed, and shows the essential inter-relationship of saltings, mudflats, and higher and lower animals.

The value of research and development

21.   The Essex Saltmarsh Research Programme has shown the complexity of issues surrounding saltmarshes and has demonstrated the need to view the system holistically. Although questions remain unanswered, the research work to date has contributed to the understanding of saltmarshes and helped to explain why saltings are disappearing. The research continues and has been recognised through its incorporation into the NRA national Saltings Research and Development programme that aims to provide practical guidance on saltmarsh management. The Essex research has already fed into the development of experimental works and flood defence management in Essex.

EXPERIMENTAL WORKS
22. In parallel with the research programme experimental works investigated practical solutions to the problems of saltings loss. This followed the instincts of the past but was radical in aiming to restore saltmarshes rather than build defences higher. The programme of works drew upon information from existing practices in Europe and the USA, and developed new approaches to foreshore management.

Polders and Groynes
23. The use of sedimentation fields is widespread in Europe, the particular approach adopted for Essex modified methods practised in Schlezwig-Holstein (Germany). A polder is created using low brushwood fences with ditches ('grips') dug to encourage sediment deposition. The aim is to increase mud levels to a point where vegetation can establish, thus regenerating saltings. Monitoring of this method demonstrated that the protection afforded by the polder led to rapid initial accretion which then slowed. Research has shown that as mud levels increase the period of tidal inundation decreases and the sediment available for deposition reduces (ref. 22). The 'gripping' of the sedimentation field creates deeper water (when flooded) thus allowing renewed deposition to take place. The mechanical compaction of sediment piled up between the ditches can attain an elevation, and stability, sufficient to support vegetation.

24. Construction of small scale polders and brushwood groynes in front of eroding saltings or flood defences have taken a variety of shapes and sizes. Monitoring of the nineteen locations around the Essex coastline shows three sites have provided no positive benefit and the remainder have variable success. Further research is needed to explain why these structures succeed in some places, and fail in others, to enable targeted use of the technique.

Wave Breaks
25. The use of structures to break down wave energy is common on beaches and this philosophy also applies to mudflats; at three sites on the exposed coast wave breaks have been constructed using redundant Thames lighters. The research indicated that protecting the profile would make it withstand storm attack more efficiently. Monitoring has shown that this approach has been successful in directly reducing the wave energy reaching the saltings, and accreting sediment on the tidal flats.

Vegetation Experiments
26. Research and experience in the USA suggested it may be possible to artificially plant vegetation to encourage the accretion of sediment or stabilise existing mudflats. Transplant experiments using Spartina (Cordgrass) and Zostera (Eel grass) on the mudflats and in polders had limited success. This approach probably failed on the mudflats because vegetation cannot survive in the (now) marginal conditions found there. Some artificial increase in elevation is needed and therefore transplants were more successful in the polder locations.

MANAGEMENT SOLUTIONS

Experiments were also carried out with false vegetation fronds. The aim was to provide an increased roughness to the mudflat, and reduce shear stress, to encourage sedimentation. Erosion of the fronds under storm conditions meant this experiment was unsuccessful.

Realignment of Flood Defences

27. To allow saltmarsh room to develop naturally the NRA carried out works at Northey Island (for English Nature) to set-back an existing defence line. The site, which is being monitored by English Nature (ref. 23), has re-established saltmarsh vegetation within two years of defence realignment. The experimental work into managed retreat by English Nature is continuing at a new location in Essex where the Ministry of Agriculture Fisheries and Food (MAFF) is conducting research into management methods, aimed at re-establishing saltmarsh and encouraging sedimentation, on previously reclaimed land. Managed retreat and the realignment of defences are options that the NRA considers when evaluating the renewal or refurbishment of any existing sea defence. Such approaches provide an opportunity to control flood water, by altering hydrodynamics, rather than trying to contain it within flood defences.

Foreshore Recharge

28. Dredgings from the Harwich Haven Authority are being used for foreshore recharge (ref. 24), sediment not used in this way is removed from the nearshore system by dumping at sea. Recharge onto mudflats has been very successful in protecting both flood defences and saltings from direct wave attack; sediments coarser than those existing at a site can maintain a steeper overall profile whilst still effectively dissipating wave energy. Clearly this alters the 'natural' system but the evidence of storm beaches on saltmarshes (ref. 10, ref. 21) suggests it is not completely alien. Monitoring of the five beach recharge sites in Essex has shown their success in protecting flood defences, and increasing the bio-diversity of the shoreline by providing habitat for such species as Nereis virens (King Ragworm). The application of this approach has now developed to a point where it is operational rather than experimental. Exploration into using coarser material continues with investigations into the potential of recharging nearshore sand banks to feed an inshore circulation system; if successful this will have long term benefits in protecting the north Essex coastline.

29. Fine sediment recharge has also been experimented with. The initial approach was to feed a limited amount of fine silt into 'L' shaped bunds, made of coarser sediment, located on a mudflat. The philosophy of this is to raise the foreshore profile, damp down wave energy, and provide sacrificial sediment for erosion. The monitoring of this work is in its infancy but the initial signs are encouraging. The second approach addresses the problem of saltings needing to grow to keep above relative sea level rise and their concomitant desire to migrate inland. Silt has

been pumped directly onto a vegetated surface, accreting an additional eight millimetres of sediment, without adversely affecting vegetation. Where saltmarshes are protected the inter-relationship of erosion and deposition (Fig.2) may make it necessary to artificially accrete sediment onto the marsh surface to enable vertical growth. At unprotected locations artificial accretion will compensate for the effects of relative sea level rise and help sustain saltmarsh vegetation. Another possible use for silts is as a part of managed retreat to raise land levels by 'warping' with layers of fine and coarse sediment to increase levels to allow vegetation establishment. It may also be feasible to raise reclaimed land to reduce the risk of flooding whilst maintaining it in agricultural production.

The value of experimental work
30. The application of research has enabled the improvement of existing methods, and development of new techniques, for flood defence. Monitoring of practical experiments has enabled the evaluation of their performance and further refinement of the methods used. Understanding traditional approaches and the errors of the past can improve management by applying the benefits of knowledge and technology available today. Adoption of the experimental work by MAFF, as good practice for coastal defence on environmental grounds, endorses this approach (ref. 25). Experimentation should continue in order to increase appreciation of the benefits in economic and environmental terms, and progress towards sustainable management methods.

FLOOD DEFENCE MANAGEMENT
The traditional approach
31. The traditional approach to flood defence (fig.3) considers only immediate frontagers and local industrial users, and has a reactionary approach to change.

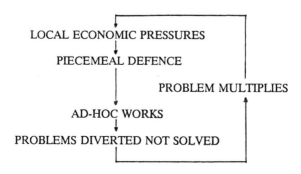

Fig. 3. Flow diagram of traditional approach to flood defence.

MANAGEMENT SOLUTIONS

This leads to piecemeal defences, using local methods of an ad-hoc nature, and has little regard for the surrounding saltmarshes, defences, coastal processes, or other coastal users. The solving of fundamental problems is not central to this approach and so they are diverted either to a later point in time or to a different location.

The Integrated approach

32. The new philosophy (fig.4) looks at wider geographical areas in terms of the defence and its' effect upon the environment; it has to cope with existing developments and relate to national interests rather than local economics. The practical application of outputs from research and development is an important element of this multi-disciplinary approach. The feedback through monitoring leads to an improved service, and expands the knowledge base, in each iterative loop. The strategic philosophy applies to all coasts but for wetlands the diversity of interests may be the greatest of any coastal type. The improvements in flood defence, achieved through the practical work on saltmarshes in Essex, is a demonstration of how the integrated management approach can work.

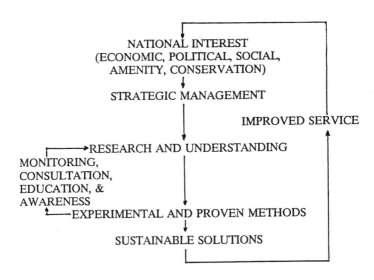

Fig. 4. Flow diagram of the integrated approach to flood defence.

33. The NRA is undertaking a strategy for the Essex coastline. The saltmarsh component of this study will be significant, with sixty per cent of the Essex flood defences fronted by them. To try to resolve the issues caused by ageing flood defences, and relative sea level rise, the NRA can

draw upon internal expertise from a number of disciplines including: flood defence, water quality, conservation, navigation, fisheries, planning advice, and water resources. Awareness of the practical realities, economics, and politics involved in providing a public service are also important in the development of a supportable strategy. In order to use this internal expertise wisely it is necessary to consult other responsible agencies and coastal interest groups. It is important to understand their perspective on the issues and, where possible, resolve areas of conflict. In the process of management it is also vital to educate the public, to enable better appreciation of their coastline, and allow open and informed debate.

The effect of economics

34. Assets of high value will be protected by flood defences, but where uneconomic a two-tier defence or managed retreat may be the appropriate solution (ref. 2). It is possible to use saltings for low density grazing and provide a buffer zone for flood defences. Existing and restored saltmarsh can be used and preserved for flood control, public amenity, and conservation. The contribution that landowners have made to the Nation, when called upon in the past to produce food, must not be overlooked. Mechanisms for providing payments for managing wetlands are needed to help change agricultural practices. Recent announcements by MAFF, on inter-tidal habitat creation and Environmentally Sensitive Area (ESA) payments, may facilitate change and support new management practices in flood defence.

CONCLUSIONS

35. We must adopt a flexible approach to management of saltmarshes. If sea level continues to rise the effects seen on saltmarshes today will persist. Planning must be for longer time periods using appropriate strategies; providing better forecasts of needs, highlighting locations that may require more (or less) expenditure, and enabling us to act with greater care towards the environment.

36. Application of the lessons of the past and the present has taken place in Essex, but the ability to look forward is only possible through using the outputs of research and applying them to practical management. We may never have definitive answers to our problems as the goal continually changes with the dynamics of the natural environment. The work on the Essex saltmarshes is an holistic approach and has challenged the way we perceive them. It has improved how we defend land based assets by using soft engineering approaches, multi-disciplinary teams, and reviewing our flood defence practices. We must leave a legacy of sustainable coastal defences, and flood control strategies, to ensure saltmarsh habitat is available for use by future generations.

ACKNOWLEDGEMENTS

The work presented in this paper has been funded by the Essex LFDC but the views expressed in it are those of the authors and not necessarily those of the LFDC, research consultants, or NRA Anglian Region.

REFERENCES

1. HR WALLINGFORD. Review of the use of saltings in coastal defence. NRA Anglian, 1988.
2. St. JOSEPH A. Environmental considerations and priorities in relation to rural seawall policies in Essex. NRA Anglian, 1992, vols. 1-2.
3. FLEMING C.A. The development of coastal engineering. Coastal Zone planning and management, Thomas Telford, London, 1992, 8-11.
4. BURD F. The saltmarsh survey of Great Britain: An inventory of British saltmarshes. NCC Peterborough, 1989.
5. CARTER R.W.G. Coastal Environments, Academic Press, London, 1988, 245-279.
6. SHENNAN I. Holocene sea-level changes and crustal movements in the North Sea region: an experiment with regional eustasy. Late Quaternary sea-level correlation and application, Kluwer Academic Publications, Dordrecht, 1989, 1-25.
7. BARRY R.G. and CHORLEY R.J. Atmosphere weather and climate. Methuen, London, 1982, 326-332.
8. PETHICK J.S. An introduction to coastal geomorphology. Edward Arnold, London, 1984, 211-233.
9. GREENSMITH J.T. and TUCKER E.V. The origin of Holocene shell deposits in the chenier plain facies of Essex (Great Britain). Marine Geology, 7, 1969, 403-425.
10. GREENSMITH J.T. and TUCKER E.V. Major Flandrian transgressive cycles, sedimentation, and palaeogeography, in the coastal zone of Essex, England. Geologie en Mijnouw, vol.55 (3-4), 1976, 131-146.
11. WILKINSON T.J. and MURPHY P. Archaeological survey of an intertidal zone: The submerged landscape of the Essex coast, England. Journal of field Archaeology, Vol.13, 1986, 177-194.
12. ROUND J.H. The Doomsday Survey, Victoria County History of Essex, 1, 369-379.
13. DIXON M. Man's effect on the coast of Essex. NRA Anglian, 1989.
14. GRAMOLT D.W. The coastal marshes of east Essex between the 17th Century and mid 19th Century. London University, 1960.
15. HMSO. Royal Commission report on flood defence in the UK, 1906-1911.
16. HMSO. Waverley committee report, 1954.

17. NRA ANGLIAN. Essex Saltings Research: 1986 to 1992. NRA Anglian, 1993.

18. PETHICK J.S. Saltmarsh geomorphology. Saltmarshes: Morphodynamics, conservation, and engineering significance, Cambridge University Press, 1992, 41-62.

19. SHEPARD F.P. and LAFOND E.C. Sand movements near the beach in relation to tides and waves. American Journal of Science, 1940, vol. 238, 272-285.

20. FREY R.W. and BASAN P.D. Coastal salt marshes. Coastal sedimentary environments. Springer-Verlag, New York, 1978, 132-143.

21. STIEVE B. and EHLERS J. The Nosse peninsula of the Island Sylt/North Frisia: Hydrology and sedimentology of storm surges. Proceedings of the International Coastal Congress, Kiel, 1992, 387-396.

22. PETHICK J.S. Long term accretion rates on tidal saltmarshes. Journal of Sedimentary Petrology, 1981, 51, 571-579.

23. IECS Hull. Set-back scheme: Northey Island tidal management, NRA Anglian, 1991.

24. POSFORD DUVIVIER. Harwich approach channel deepening: Environmental statement. HHA, 1993.

25. MAFF. Coastal Defence and the Environment: A good practice guide. MAFF, 1993.

Management initiatives for coastal wetlands

A. REYNOLDS, Environmental Scientist, and J. BROOKE, Manager, Posford Duvivier Environment

SYNOPSIS. The paper explores options for the proactive and reactive management of coastal wetlands and reviews two case studies describing coastal wetland protection and enhancement.

The paper demonstrates how the preparation of a Management Plan may provide opportunities for integrated management while enabling protection and enhancement of coastal wetlands. The valuable role of the Environmental Assessment process in identifying possible threats and determining viable mitigation and enhancement measures is also demonstrated. Overall the paper emphasises the importance of planning and positive management in ensuring the long term future of the British coastal wetland resource.

INTRODUCTION

Coastal Wetlands
1. Wetlands are defined as the habitat that occupies the transitional zone between permanently wet and generally dry environments. They share characteristics of both, but can not be classified exclusively as either terrestrial or aquatic. Within the UK a number of coastal wetlands exist. These include coastal wetlands directly influenced by the sea, for example estuarine habitats (intertidal mudflats, sandflats, saltmarsh, etc.) and those in low lying coastal areas which may not be subject to a saline influence, for example coastal grazing marsh. Both types of coastal wetlands are discussed in this paper.

Importance of Coastal Wetlands
2. The importance of coastal wetlands in Britain is demonstrated by the number of statutory designations which are present around the coast. These include Sites of Special Scientific Interest designated under the Wildlife and Countryside Act, 1981; Wetlands of International Importance

designated under the Ramsar Convention; Special Protection Areas under the EC Directive on the Conservation of Wild Birds; etc. These sites have been identified as being of importance because of their nationally and internationally important populations of birds and/or the rarity of habitats and species.

Threats to Coastal Wetlands
3. Coastal wetlands are and have been subject to many threats. A substantial decline in the total area of these habitats has been recorded. Reclamation for industrial, agricultural and development purposes has, in the past, caused a significant loss of intertidal areas. Other, more recently acknowledged loss, has been caused by coastal squeeze. Coastal squeeze occurs when development on the coast restricts the natural landward retreat of eroding coastal habitats, such as saltmarsh.
The threat of rising sea levels and any loss of sediment from the system, for example due to coast protection works, exacerbate the problem. Coastal squeeze is perhaps the most immediate threat to some coastal wetlands, leading either to the direct loss of habitat (eg. saltmarsh) or to the erosion of intertidal areas which protect sea defences in turn protecting areas of grazing marsh.

The Need for Management Initiatives
4. Considering the sensitivity of most coastal wetlands and their vulnerability to destruction and damage from development, there is a need to reach a compromise between development and conservation. Many potential threats to the nature conservation value of wetlands arise due to conflicts with other interests, for example recreation and commercial activities. Careful planning for the use of such areas through management initiatives can allow for the successful integration of interests. A variety of mechanisms are currently being used for the management of coastal wetlands in the UK.

MANAGEMENT OPTIONS

Management Objectives
5. Wetland management initiatives can be reactive (a response to an particular threat) or pro-active (planned to avoid future conflicts). Such management can have different objectives. For example it can be directed towards preservation, conservation, or enhancement of the resource. In some cases, different objectives will be appropriate in different parts of a coastal wetland site. These management objectives are defined in Table 1 below.

Table 1 Wetland Management Objectives

Term	Definition
Preservation	The maintenance of the resource in exactly the same state as at present, irrespective of any natural or human-induced change.
Conservation	The protection of the resource whilst permitting its natural (or beneficial) development (eg. the transgression from mudflat to saltmarsh).
Enhancement	Intervention to increase the value of the resource in terms of its area, diversity, sustainability, etc. Enhancement can relate to improving the quality of an existing resource (on site) or creating or re-creating an additional area of that resource in an adjacent or removed location.

Scale of Management

6. Management initiatives, as well as varying in their objectives or purpose, can vary in scale. Management may be required for a specific site or system, or for part of that site. On a larger scale, the overall resource may require maintenance or protection at a more strategic level. This scale of management may relate to an estuary, a defined length of coast, or even the national resource.

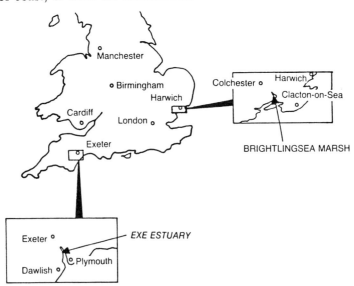

Figure 1 Location of Case Studies

Proactive Management

7. Pro-active management initiatives in the UK at the present time include the development of (non-statutory) shoreline, estuary or coastal Management Plans. These management initiatives and their application to the management of British coastal wetlands are defined in Table 2.

Table 2 Current Management Planning Initiatives

Type of Plan	Purpose of Plan	Relevance to Coastal Wetlands
Shoreline Management Plan	Plan dealing with the long term management of a defined length of coast for coastal defence purposes. Other interests (eg. conservation, archaeology, etc) taken into account.	Many coastal wetlands play an important role in coastal defence. Saltmarshes, for example, act as a buffer absorbing wave energy and hence reducing erosion of the sea wall. In other cases, a coastal defence structure or scheme might be required to protect freshwater coastal wetlands against saline inundation.
Estuary Management Plan	Integrated plan covering management of all interests, both on land and water areas. Estuary Management Plan defines actions necessary to resolve current problems and defines policies to deal with possible future conflicts. Relevant policies may feed into statutory local plans.	Protection for coastal wetlands, or initiatives for their enhancement, may be defined in Estuary Management Plan. These can be direct (eg. planting, or the prevention of access) or indirect measures (eg. zoning an activity to an alternative location; providing interpretation facilities; etc). In addition, policies can be defined to guide future decision making on issues that might affect coastal wetlands.
Coastal Zone Management Plan	As above for Estuary Management Plan, but for a defined length of coastline (defined by physical or administrative limits).	Generally as for Estuary Management Plans.

MANAGEMENT SOLUTIONS

Reactive Management
8. Reactive management needs are commonly, although not always, identified as a result of undertaking strategic studies into the status of a resource, or Environmental Assessments or similar in respect of a proposed change or development. The environmental assessment process, in particular the identification and evaluation of the impacts on existing environmental characteristics, may highlight recent or anticipated future trends. The Environment Assessment process will also highlight potentially beneficial or adverse impacts where enhancement or mitigation would be possible (or necessary).

9. In some cases, mitigation or enhancement may comprise one-off measures (eg. replacement or establishment of a species, creation of a habitat, etc.). In others, mitigation or enhancement may take the form of long term management (eg. by facilitating the control of water levels; by maintaining protection against saline inundation of freshwater wetlands; by initiating monitoring, etc.).

Case Studies
10. The following sections describe two management initiatives for British coastal wetlands. The first case study describes the way in which coastal wetland protection and enhancement was incorporated into the Exe Estuary Management Plan, Devon. The second case study describes the management implications of a proposed tidal defence scheme for a coastal wetland at Brightlingsea, in Essex. The location of these case study sites is shown on Figure 1.

MANAGEMENT CASE STUDIES
Exe Estuary Management Plan

Introduction
11. As indicated in Table 2, coastal wetland management can be achieved via the implementation of Management Plans, such as those currently being prepared for several British estuaries. Many of the UK's major wetland habitats are included within estuarine ecosystems. The majority of estuaries comprise a mosaic of wetland habitats including mudflats, sandflats, saltmarsh, reedbeds, etc. together with grazing marsh habitats on the surrounding low lying land. The importance of the UK's estuaries in terms of a wetland resource is reflected in the designation of many estuaries as Wetlands of International Importance under the Ramsar Convention. The Ramsar Convention promotes the "wise use" of wetlands through "their sustainable utilisation for the benefit of human kind in a way compatible with the maintenance of the natural properties of the ecosystem".

Background

12. The Exe Estuary in Devon is typical of this type of ecosystem. It is internationally and nationally important, being designated as a Ramsar Site, a Special Protection Area and a Site of Special Scientific Interest. The estuarine ecosystem contains a variety of coastal wetlands with extensive mudflats and sandflats, areas of saltmarsh and, on the reclaimed floodplains, large areas of grazing marsh. However, the estuary hosts a wide range of other interests. It is a regionally important site for recreation and an important base for commercial users including fisheries and tourist-based industries. Considering the variety of interests on the Estuary it is surprising that current conflicts between users are relatively infrequent. However, for the Exe Estuary to remain in this state it is important to ensure that proactive management is implemented through a management initiative. Posford Duvivier Environment was therefore commissioned to prepare a draft Management Plan for the Estuary.

The Need for Management

13. Given the importance of the Exe Estuary in terms of its nature conservation significance and its value for tourism, recreation and commercial uses, it is essential to ensure that all interests are represented in its management. The present fragmented and uncoordinated actions undertaken by a range of statutory and non-statutory bodies could potentially lead to increased conflicts between nature conservation and other interests. Although not an immediate issue, there is a potential threat that developing conflicts may cause the degradation or loss of the internationally important wetlands. For example, the unregulated increase in the numbers of windsurfers and jetskiers using the Estuary may cause disturbance to feeding birds and damage to the intertidal habitats.

Goals of Management

14. The draft Management Plan defined the overall goal of management of the Exe Estuary thus:- "to promote the sustainable use of the Estuary to yield the greatest benefit to the present population whilst maintaining its potential to meet the needs and aspirations of future generations, in a manner compatible with the maintainance of the natural properties of the Estuary and its value for wildlife."

Management Terminology

15. In order to achieve this goal, a series of objectives, policies and actions were identified. These terms are defined in Table 3. The flow chart at Figure 2 shows the relationships between the various elements of the management planning process

MANAGEMENT SOLUTIONS

Table 3 Definition of Terms Used in the Management Plan

Term	Definition
Goal	Broad, long term aim, or vision, of the Management Plan. Goals encompass the holistic management of the estuarine resource.
Objective	Specific, shorter term target for attaining goals. Objectives may focus on specific environmental parameters or on achieving a particular goal.
Policy	Statement setting out agreed guidance on a particular issue or range of issues. Policies provide the framework within which future decisions will be taken.
Action	Detail on what needs to be done, by whom, and when, in order to achieve a particular objective (or an element of that objective).

<u>Objectives of Management</u>

16. Amongst a whole series of objectives developed for the management of moorings, navigation, fisheries, and many others, those key objectives which directly or indirectly relate to the coastal wetlands are shown in Table 4.

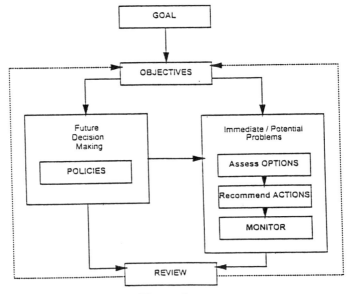

Figure 2
The Management Planning Process

Table 4   Selected Exe Estuary Management Objectives

Topic	Objective
Moorings	Agree a plan to ensure that mooring numbers are controlled and that moorings are managed to minimise conflict with other users.
Water Based Recreation	Provide for the continuing and safe use of the Estuary by recreational watercraft while ensuring that ecological damage is prevented and that conflict with other activities is minimised.
Land Based Recreation	Enable land based and intertidal recreational activities to take place without damaging the estuarine resource or disturbing other users.
Natural Environment	Identify opportunities for maintaining and improving the nature conservation value of the Estuary and its environs whilst ensuring that conflict with other users and activities is minimised.
Research	Set up a comprehensive and targeted monitoring and research programme in order to measure and improve the effectiveness of management.

Management Policies

17. The Management Plan then identifies a series of policies designed to guide future decision making to ensure that the overall objectives are achieved. In terms of the coastal wetland interests on the Estuary, the following policies are of particular relevance:-

- To provide for windsurfing in the Estuary whilst ensuring no undue conflict with other interests.
- To ensure that moorings do not encroach on sites which are of demonstrated importance to other interests.
- To ensure that bait digging is carried out on a controlled basis, at a sustainable level.
- To control and regulate wildfowling activities on the Estuary.
- To restrict disturbance from loose dogs.
- To maintain and enhance the nature conservation value of the Estuary.
- To reduce disturbance to wildlife.
- To develop an improved understanding of the estuarine resources.
- To promote and participate in appropriate research projects.

MANAGEMENT SOLUTIONS

Management Actions

18. Finally, in order to deal with existing or anticipated short term problems, a series of actions are defined in the Exe Estuary Management Plan. Actions which require control or change on the part of other interests, but which are designed to protect the wetland resource include:-

- Monitor the effect of windsurfing on *zostera* beds and feeding birds, and review results through a working group to facilitate discussion. Depending on results, possibly create formalised windsurfing zones.
- Using a harbour revision order, confer licensing powers on the harbour authority to control all mooring by granting licences to moor and thereby limit the number and location of moorings. This would enable moorings to be controlled in the interests of nature conservation by avoiding sensitive habitats.
- Set up a working party involving wildfowling and nature conservation interests to discuss the shooting requirements within the Estuary. Re-evaluate the areas presently shot over, and assess the disturbance to birds resulting from night shooting. This would involve a reassessment of the bird sanctuary order, possibly leading to the removal of the order from less sensitive areas, and increased restrictions on shooting in sensitive areas.
- Control bait digging through licensing of commercial operators, a high profile education and information programme, the formation of a local bait diggers association, the designation of a bait digging exclusion zone, and the monitoring of any disturbance to birds.

Actions for Enhancement

19. In addition to those actions requiring the management of other interests, several actions were defined to help enhance the natural estuarine resources. These include:-

- Identify opportunities for habitat creation and enhancement. The most obvious area in terms of coastal wetlands, is Exminster Marshes where land purchase and an improved flood regime would have positive benefits for the nature conservation value. However, it is important to stress that priority should be given to the maintenance of the present nature conservation value of the Estuary, through the prevention of disturbance to, and destruction of, the present habitats.

- Improve information on the nature conservation value of the Estuary. Both the Estuary and the public would benefit from improved interpretation facilities at access points and car parks. This information should explain the importance of the Estuary in terms of nature conservation, together with the relationships between the various uses.

20. As indicated above, the Management Plan highlighted the opportunity to improve the nature conservation value of Exminster Marshes. Exminster Marshes are important as a grazing marsh habitat. They also form part of the National River Authority's flood alleviation strategy for Exeter and lower Exe valley. The water levels in the ditches are therefore maintained at a level appropriate to their flood alleviation role. The grazing marshes could, however, be improved for nature conservation purposes if the water levels in the ditches were raised, and if more regular shallow winter flooding of the marshes was encouraged. The Management Plan recommended the establishment of a formal working party to ensure the continuation of discussions between the involved agencies.

Summary
21. The draft Management Plan for the Exe Estuary therefore sets out a relatively formal, albeit non-statutory, framework for the management of the area's coastal wetlands, integrated with the needs of the multitude of other users and interests. Public consultation on the draft Management Plan may lead to some changes or additions. Given the agreement of all parties, however, it is intended that the Plan will facilitate the protection and enhancement of the wetland resource through positive policies, some of which may subsequently be adopted as statutory policies in the relevant local authoritys' Local Plans.

BRIGHTLINGSEA TOWN TIDAL DEFENCES ENVIRONMENTAL ASSESSMENT

Introduction
22. The defences to the town of Brightlingsea are below the appropriate standard for a built up area. The National Rivers Authority's preferred option is to improve the existing defences and to construct new ones along the low lying quayside which is not currently protected. A secondary defence will be constructed where required to protect houses and property. The Environmental Assessment examined a wide range of options and enabled the NRA to make an objective decision on the environmental aspects of the preferred option.

MANAGEMENT SOLUTIONS

23. As part of the scheme development, the National Rivers Authority commissioned Posford Duvivier Environment to undertake an Environmental Assessment under the Town and County Planning (Assessment of Environmental Effects) Regulations 1988, to ensure that all potential environmental impacts were identified and that appropriate mitigation measures were determined.

24. The Environmental Assessment investigations covered all aspects of the natural, human, physical and chemical environment. The following discussion concentrates on the role of the Environmental Assessment in addressing nature conservation issues. It must be stressed, however, that all environmental issues not just those pertaining to nature conservation were taken into account when determining the preferred option. It should also be noted that the final outcome of the various studies, including the Environmental Assessment, is not yet determined. The current position is that the National Rivers Authority have identified their preferred option and are seeking to promote it.

Background
25. Brightlingsea Marsh is acknowledged as an important site in terms of its nature conservation value. It has been included in the Colne Estuary Site of Special Scientific Interest, Ramsar Site and Special Protection Area. That part of the site which is managed by English Nature has also been designated as a National Nature Reserve. The predominantly freshwater marsh provides a unique habitat of tussocky unimproved swards and associated ditch systems. The site is noted for its high botanical species diversity, freshwater invertebrate interest, and ornithological importance, the latter both for overwintering and breeding birds. The vulnerability of this type of habitat is demonstrated by the substantial decline of grazing marsh habitats over the last 50 years. For example, between 1938 and 1981, 82% of grazing marsh in Essex was lost, primarily to agricultural intensification.

Existing Management
26. English Nature presently manage the marsh through leases and management agreements. Water level controls and management through careful grazing are used both to ensure that the present ecological importance is retained, and to implement opportunities for enhancement. In particular, recent improvements in the management of water levels has encouraged the use of the marsh by overwintering wildfowl and waders. The current presence of the intact flood defences also makes possible the management of the site as a largely freshwater resource.

Ecological Impact Investigations

27. Investigations undertaken as part of the Environmental Assessment demonstrated that any abandonment of the existing flood embankment which protects the marsh would have detrimental consequences, eventually leading to a deterioration in the ecological value of the grazing marsh habitat. The ecological characteristics of the marsh have developed under a predominantly freshwater regime and a lack of excessive saline influence. Increased overtopping, or more detrimentally a breach in the embankment, would cause a change from freshwater and oligohaline (brackish) conditions to mesohaline (saline) conditions. A literature search and review was undertaken in order to assess the likely impacts of saline inundation and flooding on the plant, invertebrate and ornithological communities. This concluded that significant and prolonged increases in salinity and inundation would be likely to result in a loss of many of the botanical, invertebrate and ornithological communities which presently occur on the marsh.

Management Impacts

28. Any deterioration in the condition of the flood defences leading to increased overtopping of saline water, or a breach leading to the regular or permanent inundation of parts of the site, would similarly have detrimental consequences for the management of the site. High salinity levels could eventually make grazing difficult or impossible. A breach could lead to significantly changed drainage/hydrological regime. Both would hinder, and eventually prevent, the continuation of the existing management regime at the site.

Mitigation and Management Opportunities

29. Given the site's importance reflected by its Ramsar, SPA, SSSI and NNR status, the potential ecological implications of abandoning the defence were consequently regarded as significant. In addition to the proposed retired line of defences protecting the town (see Figure 3), it was therefore recommended that the present defence around the marsh should be upgraded and subsequently maintained. Such an improvement would mean that even a 1 in 100 year flood event would be unlikely to significantly impact the marsh, having a shallow depth and short duration. The increased security associated with the reduced risk of flooding would mean both that the existing management of the site would be able to continue, and that investment in future opportunities for improved management (eg. further improvements in water level control to provide better conditions for overwintering wildfowl and waders) could be made in confidence.

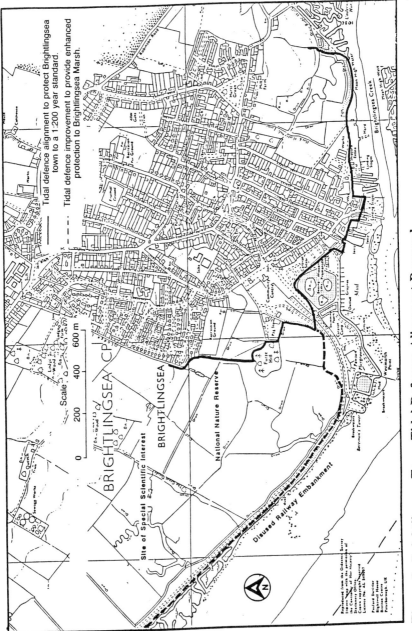

Figure 3 Brightlingsea Town Tidal Defences Alignment Proposal

Summary

30. Part of the Environmental Assessment process therefore identified an ongoing management initiative for nature conservation purposes which might be threatened by a decision in respect of flood defence provision. The mitigation measures recommended, however, not only safeguarded the management of existing interests at the site, but also offered an improved standard of protection with associated opportunities for enhanced ecological management.

Conclusions

31. This paper identifies various management options currently being applied to, or investigated for, British coastal wetlands. The case studies provide examples of both proactive and reactive management. The key lessons which can be drawn from these case studies to be applied more generally to the management of coastal wetlands include:-

- The preparation and implementation of a plan for integrated management, may provide both protection and enhancement to coastal wetlands whilst also meeting the requirements of other interests.
- Environmental Assessment has a valuable role, not only in identifying possible threats to coastal wetlands, but also in determining viable mitigation measures and possible enhancement opportunities.
- An increased emphasis on careful planning and positive management can help to ensure the long term future of the British coastal wetland resource.

REFERENCES

1. Posford Duvivier Environment. Brightlingsea Town Tidal Defences: Environmental Statement. 1993. Posford Duvivier Environment, Peterborough. PE3 8DW.

2. Posford Duvivier Environment. The Future Management of the Exe Estuary: Management Plan. 1993. Posford Duvivier Environment, Peterborough. PE3 8DW

3. English Nature. Estuary Management Plans - A Co-ordinators Guide. 1993. English Nature, Peterborough.

Seasonal changes of groundwater chemistry in Miyatoko mire

Dr T. HIRATA, Dr S. NOHARA and Dr T. IWAKUMA, National Institute for Environmental Studies, C. TANG, Graduate School of Science and Technology, Chiba University, and Dr K. NAKATSUJI Dept of Civil Engineering, Osaka University, Japan

SYNOPSIS The chemistries of the ground and surface waters in the Miyatoko mire of Japan have been monitored since 1991. The groundwater chemistry has the typical features of a peat land characterized by high NH_4-N and SiO_2 concentrations, a long residence time, and an anaerobic environment. The chemistry of the surface water, originating from spring located at the skirt of surrounding mountain and running through the wetland, demonstrates that pH and the SiO_2, Na and K concentrations tend to show lower values at the outlets of the mire rather than the springs. This result can be explained by the runoff of pool water with its short contact time with the top soil and plants.

INTRODUCTION

1. Groundwater is basically responsible for adjusting the drainage of water from the catchment and is also closely related to the soil water and its evapo-transpiration from the ground surface. Groundwater in mires, as a particular form of groundwater, is characterized by the water table close to the ground surface, with the flow and residence properties of wetland groundwater playing a major role in the upkeep of the wetland ecosystem. In other words, the drastic changes in groundwater flow and soil water due to catchment conversion and changes in land use may give rise to changes in the species inhabiting the wetland biotope with the unacceptable prospect of the wetland itself becoming extinct.

2. Groundwater has a markedly slower flow than surface water, and the changes in groundwater chemistry associated with time and space are thus capable of providing information indicative of the bio-chemical reaction of the groundwater to the patterns of land use and the downgradient flow processes. Compared with the properties of surface water, it is clear that the drainage and residence behavior of groundwater may also serve as a better indicator of the water cycle in the catchment. In view of these features, the behavior of groundwater can shed light on the distribution and transition patterns of the biological species in the wetland. It can also provide valuable insights into the mechanism by which the wetland is maintained in a stable and steady balance. The field monitoring was conducted in order to investigate the role groundwater plays in the upkeep of the wetland and consisted of a survey of the groundwater and surface water in the Miyatoko mire located in Minami Aizu, Fukushima Prefecture Japan. The paper describes some of the major findings of chemical changes with migration of ground and surface waters in the Miyatoko mire.

STUDY SITE DESCRIPTION AND MEASUREMENTS

3. The Miyatoko wetland is located at latitude 38°N and longitude 140°E (Minami Aizu, Fukushima Prefecture), at an elevation of approximately 850m above sea level. Stretching over a surface area of 6ha, it is completely ringed by mountains. Its catchment area, determined from the water divide on the topographical map, is 54.1ha.

4. Fig. 1 shows a topographical map of the wetland drawn with 10cm elevation intervals. Along the longitudinal axis, the wetland has a near-elliptical plan view measuring approximately 400m across the major (longitudinal) axis and approximately 170m across the minor (lateral) axis. Fig. 1 shows the relative differences in elevation with respect to a certain point in the wetland taken as the reference point. The topography of the site generally inclines from the northwest towards the southeast. Along the 400m in the wetland's longitudinal axis, there is a gradient of 2.2m.

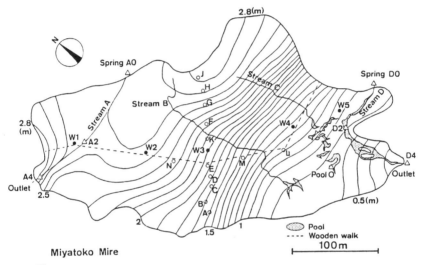

Fig. 1 Geographical map and measurement locations in the Miyatoko mire

5. Skirting the periphery of the wetland are the surrounding mountains, evidencing groundwater effusion. The Koshimizu (A0) and Oshimizu (D0) springs are the most prominent water sources, both of which flow through the wetland. These two springs rise at opposite ends of the wetland. In addition to two streams, there are two more water flows in the wetland, described as B and C, respectively. These two flows carry water after the thaw in March and April through the wet seasons in July and September.

6. To study the groundwater flow in the wetland, a wooden path laid along the longitudinal axis of the wetland zone was used to establish a total of five locations for observation. These consisted of the four points from W1 through W4 and a fifth point W5 taken as the seepage location of groundwater to stream D in the longitudinal direction. At point W1, observations were carried out in PVC investigation wells (with 2cm internal diameters) at two depths, i.e., 0.5m and 1.5m, from the ground surface. At the remaining four points from W2 through W5, the measurements took

place at three depths, i.e., 0.5, 1.5 and 2.5m. Apart from these five observation points spaced out along the longitudinal direction, there were also nine measurement points from A through J across the center line of the wetland to examine the groundwater flow in the lateral direction. To correct the measurements performed in the longitudinal direction, three further measurement points were added, described as L, M, and N, respectively. These twelve investigation wells had a depth of 1m.

7. Except for the winter months, with its heavy snowfalls from January through March, site observations have been continued with one measurement per month from 1991. The groundwater levels have been measured on each occasion at the five longitudinal measurement points from W1 through W5 and the 12 additional observation points. Groundwater samples have been taken at the five longitudinal measurement points from W1 through W5. The surface water has been sampled on each occasion at the Koshimizu Spring (A0) and the outlet point A4 for the water of the stream A, and at the Oshimizu Spring (D0), pool (D2) in the middle of the waterway and the outlet point D4 for the stream D.

GROUNDWATER FLOW SYSTEM

8. Fig. 2 shows the seasonal changes in the groundwater's total potential from May 1991 through November 1992. The result illustrates the height or total water head, in cm, of the groundwater level at each of the investigation wells measured from the reference point. In this study, the reference point for the potential is exactly the same to the reference point used for the geographical map. When the groundwater's total potential is found to be identical with the topographical elevation, the groundwater level is to appear at the ground surface. Fig. 2 also gives the monthly rainfall data. From 1990 through 1993, annual rainfall (during the period from January through December) is given as 1219, 1628, 1310, and 1440mm, respectively, giving an average of 1399mm. The potential evapo-transpiration was calculated using the Hamon's equation (ref. 1) for the years 1991 and 1992 during which the analyses were performed. The results of these calculations are 714 and 680mm, respectively. The effective precipitation contributing to water outflow was 914mm in 1991 and 630mm in 1992.

9. Comparison of the potential differences for the different investigation wells shows that there was practically no difference in 1991 at the 50cm depth of W1 and W2. In 1992, however, W2 had a higher potential. At the deeper depth of 150cm, W2 had a higher potential in 1991, though only by a very small margin. Conversely, in 1992, W1 had a higher potential. This shows that the groundwater does not have a uniform or constant flow pattern year in, year out. The groundwater at W1 and W2 near the surface flows generally towards the northwestern edge of the wetland.

10. The total potential at the 150cm and 250cm depths decreased in the order from W2 through W3 and W4 to W5 in 1991. The groundwater in these locations is believed to flow towards the southeastern rim of the wetland along the geographical inclination of the land in the longitudinal direction. In 1992, however, a significant drop in total water head was measured at the 150cm and 250cm depths at observation point W3. At the measurement depth of 50cm of wells W4 and W5, the total potential marked a reverse movement. It may be acceptable to suppose that generally, the predominant flow pattern is along the geographical inclination of the land.

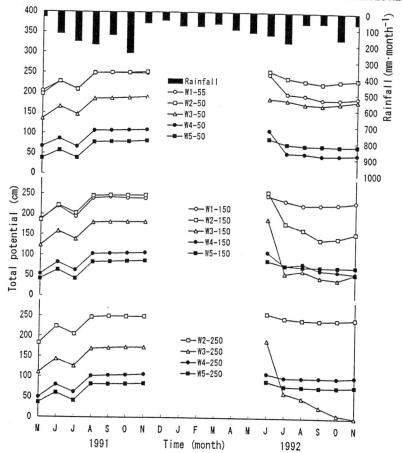

Fig. 2 Seasonal change of total potential of groundwater in longitudinal axis of the Miyatoko mire. The values following the well number of W1 through W5 denote the well depth in cm.

11. The measurements of the total water head at the investigation wells from A through J (see Fig. 1) showed that the total potential tended to fall from A through B to C, in this order, from the west and in the order from J through H, G, F, K and E to C from the east. These results indicate that the groundwater supplied from the surrounding mountains at the eastern and western flanks of the wetland collects at points B and C. Similarly, it was found that the total potential in the measurement locations L, M and N, established to correct the measurement in the longitudinal direction, dropped in the order from N through M to L. This suggests that the longitudinal flow towards the southeastern side of the wetland is predominant. The results suggest that while there is a certain groundwater flow component following stream A in some parts of the northwest, the remaining majority of the groundwater flows in the longitudinal direction, that is, the predominant direction consistent with the geographical inclination of the land, to flow out in the southeast edge.

GROUNDWATER CHEMISTRY

12. Groundwater has a very sluggish flow and so it normally does not show significant changes in its properties, which tend to remain stable. As can be seen in Fig. 3, the seasonal changes in groundwater pH in the Miyatoko wetland exhibit a certain regular pattern of peaking in the summer season and declining from the autumn to the winter. A similar pattern of change can be found for Na, the concentration of which tends to increase from the spring to the summer and decrease in the winter. It was not possible, however, to discover any systematic pattern of seasonal change for

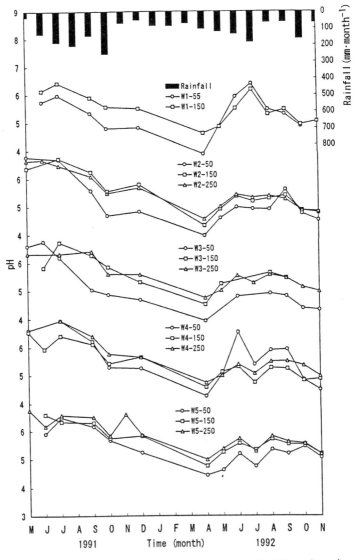

Fig. 3 Seasonal change of pH in groundwater in the Miyatoko mire

the other chemical substances. Table 1 gives the annual mean values for solute concentrations in the wetland's groundwater for the period from 1991 through 1992.

13 To analyze the groundwater flow it is normally inevitable to establish measurement points in the flow direction of the groundwater. Yet, the construction of an elaborate observation system for precise measurement bears the risk of causing damage to the plant life in the wetland. For the study, a wooden path running through the wetland was therefore used as a less intrusive alternative for conducting the measurements. This means that the investigation wells from W1 through W5 do not necessarily lie in the same groundwater flow. This system is nevertheless adequate to permit the detection of some of the specific features of the wetland. First, it has been found that the groundwater has a significantly lower NO_3-N concentration than ordinary surface water (ref. 2). In contrast, however, its NH_4-N and SiO_2 concentrations tend to be higher. The NH_4-N concentration was monitored for two years at the wells from W2 through W4, with the annual mean concentration being in excess of 1mgl^{-1}. Similarly, many of the observation points demonstrated SiO_2 concentrations at annual mean levels exceeding 30mgl^{-1}. SiO_2 originates from the soil so the concentration in the groundwater will increase with increasing the contact time between the

Table 1 Annual mean values of solute concentration of groundwater in the Miyatoko mire during two-year period of 1991 to 1992

	NH_4-N	NO_3-N	PO_4-P	SiO_2	Cl	SO_4	Na	K	Ca	Mg	pH	SC
Apr.-Dec. 1991 (mg·l^{-1})												(μS·cm^{-1})
W1- 55	0.330	0.008	0.11	9.87	5.7	5.65	2.04	3.95	2.07	0.95	5.32	29.2
150	0.282	0.005	0.027	21.1	11.2	4.18	2.29	8.14	2.15	1.44	5.90	68.7
W2- 50	1.32	0.005	0.14	28.1	12.4	9.67	1.87	9.63	4.67	2.50	5.71	92.1
150	2.28	0.003	0.067	30.6	4.62	7.03	1.68	2.43	5.32	2.97	6.13	60.4
250	2.93	0.008	0.044	32.5	3.85	5.53	1.77	2.25	4.68	2.39	6.17	65.5
W3- 50	1.49	0.001	0.065	14.6	5.32	9.6	1.79	2.23	4.07	2.05	5.68	45.4
150	4.00	0.011	0.13	37.8	5.34	13.1	2.34	1.71	5.7	3.45	6.0	66.6
250	4.21	0.005	0.075	42.9	4.05	9.07	2.35	1.97	5.38	2.96	6.05	66.2
W4- 50	1.01	0.006	0.056	23.8	4.15	7.02	3.0	3.11	4.57	2.19	6.06	44.8
150	2.073	0.001	0.092	37.0	4.43	5.34	2.86	3.18	2.7	1.49	6.01	47.0
250	2.70	0.008	0.064	37.1	3.82	2.94	2.71	2.68	2.32	1.46	6.20	38.9
W5- 50	0.51	0.008	0.157	30.6	4.1	6.12	4.04	2.47	2.97	1.55	5.89	27.6
150	0.34	0.009	0.026	32.3	2.8	2.68	4.09	2.03	2.77	1.51	6.16	31.2
250	0.256	0.005	0.035	30.3	3.2	3.3	4.29	1.98	3.69	2.05	6.33	45.5
Apr.-Dec. 1992 (mg·l^{-1})												
W1- 55	3.50	0.026	0.593	10.1	6.22	6.33	2.47	4.02	1.88	0.57	5.30	27.6
150	0.454	0.025	0.022	22.3	2.06	4.24	1.74	0.729	1.79	1.18	5.25	21.7
W2- 50	2.902	0.007	0.132	22.8	3.83	7.26	1.66	1.30	1.06	0.75	4.81	23.2
150	3.43	0.005	0.075	25.2	4.1	7.1	1.94	1.55	0.7	0.50	5.07	26.8
250	3.76	0.005	0.047	27.8	3.04	7.17	1.70	0.971	0.97	0.65	5.11	29.8
W3- 50	2.43	0.005	0.059	13.8	5.1	10.4	1.76	0.885	1.38	0.84	4.56	20.9
150	5.23	0.018	0.185	37.1	6.8	14.8	2.75	2.03	1.33	0.86	5.22	40.1
250	6.20	0.007	0.160	42.2	4.44	9.99	2.80	2.30	1.03	0.62	5.23	41.4
W4- 50	1.97	0.012	0.039	24.0	2.37	5.5	2.12	0.893	0.73	0.52	4.72	22.4
150	2.73	0.004	0.038	37.1	1.81	4.94	2.48	0.870	0.79	0.49	5.04	26.5
250	3.28	0.010	0.011	35.8	1.6	2.29	2.19	0.726	0.74	0.59	5.25	30.1
W5- 50	0.570	0.006	0.048	35.8	2.09	4.84	2.9	0.865	1.31	0.78	5.06	19.7
150	0.670	0.008	0.006	32.8	1.94	1.93	3.44	0.579	2.47	1.45	5.42	27.9
250	0.543	0.004	0.007	30.5	1.94	2.33	3.36	0.643	2.79	1.59	5.52	28.5

water and the soil. This leads to the inference that the groundwater in the wetland has a fairly long residence time. The analyses also showed that NO_3-N was practically absent while NH_4-N raised the concentration, suggesting that the groundwater in the Miyatoko wetland has the typical anaerobic characteristic of a peat land.

14. Fig. 4 shows the areal distribution of the NH_4-N and SiO_2 concentrations in the Miyatoko groundwater, using the 1991 measurements. The concentrations at the wells from W1 through W5 are the measurement data obtained at the depths of 50, 150, and 250cm from the left for all measurement points. Since the contact time between the water and the soil is generally longer at deeper depths, at most observation points the NH_4-N and SiO_2 concentrations were found to be higher at deeper groundwater. The SiO_2 concentration at W1 is recognized to be lower than at the other four wells so that the groundwater has a shorter residence time in this location. The NH_4-N concentration was also lower at W1 rather than the other wells W2 through W4. These features and the fact that the groundwater in the western part of the wetland flows along the stream A may be taken as evidence substantiating the different flow dynamics and residence behavior of this groundwater from the major groundwater flowing in the longitudinal direction.

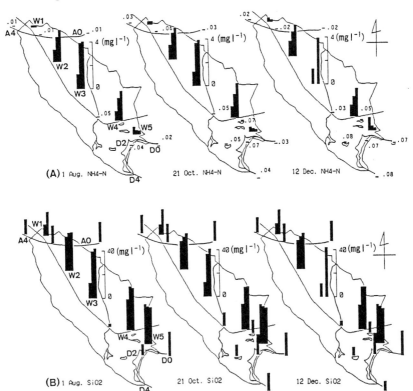

Fig. 4 Horizontal distribution of NH_4-N (A) and SiO_2 (B) concentrations in the Miyatoko mire during a period of 1991

SURFACE WATER CHEMISTRY

15. The wetland has four streams. Table 2 lists the annual mean values for the downgradient changes in the chemistry of the surface water for the two main streams among them, that is, streams A and D. Fig. 5 shows the downgradient seasonal changes of the surface water chemistry. In terms of the pH value, it was not possible to detect any substantial differences between the spring and outlet points in the annual mean values for both surface waters. In response to the higher rainfall from June through August 1991, the water at the Koshimizu and Oshimizu outlets reduced in pH which then rose again as winter approached. After the winter, the spring thaw again showed an increase in pH.

16. The SiO_2 concentrations at the Koshimizu and Oshimizu springs were monitored once a month. Within this limitation, the results remained consistent throughout the observation period. The mean annual concentration was 20.5mgl^{-1} in 1991 and 20.7mgl^{-1} in 1992 at the Oshimizu spring D0. These values are not substantially different from the mean concentration levels for rivers from the Kanto region to Tohoku which are in the range from 21.5 - 23.1mgl^{-1} (ref. 3). The SiO_2 concentration at the Koshimizu spring A0 is lower than at the Oshimizu spring D0, but it showed no seasonal variations. In contrast, the concentration at the wetland outlet is lower than at either of the springs. Similarly, the seasonal changes presented a pattern close to that for the pH. Both at the spring locations and the wetland outlets, the solutes Na and K showed a behavior similar to that of SiO_2.

17. The NO_3-N concentration emerged as a characteristic feature of the Miyatoko wetland. As mentioned for the wetland groundwater, there was practically no evidence for the presence of NO_3-N in the spring water and spring outlets. The NH_4-N, Ca and Mg concentrations showed a minor increase at the outlets as compared with their concentrations in the spring water. This difference was so small as to be negligible.

Table 2 Annual mean values of solute concentration of surface water in the Miyatoko mire during two-year period of 1991 to 1992

	NH_4-N	NO_3-N	PO_4-P	SiO_2	Cl	SO_4	Na	K	Ca	Mg	pH	SC
Apr.-Dec. 1991 (mg·l^{-1})												(μS·cm^{-1})
Spring D0	0.034	0.006	0.008	20.5	2.91	0.86	2.56	2.03	0.622	0.28	6.52	37.8
Pool D2	0.070	0.002	0.028	14.7	3.11	1.8	2.15	2.11	0.643	0.317	6.25	28.1
Outlet D4	0.077	0.009	0.028	12.7	4.33	2.05	1.95	2.61	0.675	0.389	6.11	36.8
Spring A0	0.036	0.006	0.009	16.9	2.81	1.08	2.33	1.71	0.533	0.268	6.35	32.3
Outlet A4	0.033	0.012	0.009	14.0	3.25	1.31	2.15	2.23	0.5	0.334	6.22	27.8
Apr.-Dec. 1992 (mg·l^{-1})												
Spring D0	0.015	0.005	0.001	20.7	2.18	0.86	2.36	1.21	0.55	0.289	5.97	16.3
Pool D2	0.045	0.006	0.001	18.3	2.39	1.33	2.12	1.04	0.61	0.278	6.10	15.1
Outlet D4	0.077	0.009	0.028	12.7	4.33	2.05	1.93	2.61	0.68	0.389	6.11	36.8
Spring A0	0.026	0.008	0.002	17.1	2.31	2.69	2.13	0.95	0.56	0.272	6.09	14.6
Outlet A4	0.017	0.013	0.002	14.4	2.61	1.37	1.85	0.73	0.53	0.29	6.11	16.0
Apr.-Dec. 1991 (mg·l^{-1})												
Pool 0	0.072	0.002	0.006	8.5	3.88	2.78	0.882	0.464	0.425	0.33	5.20	25.9

MANAGEMENT SOLUTIONS

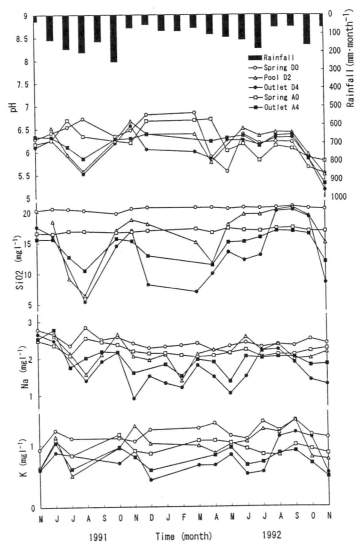

Fig. 5 Downgradient chemical change of surface water in the Miyatoko mire

DISCUSSION AND SUMMARY

18. The pH and SiO_2 concentration of surface water show a relatively clearly defined pattern of seasonal changes. The changes may be attributed to the effect of the runoff of groundwater and rainwater. The runoff of groundwater was examined by carrying out a survey of the groundwater chemistry in the Miyatoko wetland at depths below 50cm for five measurement points located essentially along the longitudinal axis. Of these five measuring points, W5 is the closest to stream A, yet the outlet of the Oshimizu spring D0 has a height of about 0.4m and the pool depth on the

downgradient is also 0.5m. Moreover, the elevation of well W5 is 0.85m, so that it is significantly downstream of pool D2 and spring outlet D4 even though the groundwater at the depths of 150cm and 250cm in investigation well W5 does flow out. This suggests that the groundwater runoff of W5 has no direct impact on the solute concentrations in the pool and outlet for the stream D. At the depth of 50cm in W5, the annual mean concentrations for NH_4-N and SiO_2 exceed 0.5mgl^{-1} and 30mgl^{-1}, respectively. Both concentrations are higher at the Oshimizu spring. Na and K are also generally higher at the springs. This suggests that seasonal changes in solute concentrations cannot be explained in terms of the runoff of this type of groundwater with these properties.

19. The seasonal changes in SiO_2 and Na measured in the surface water can be explained, however, if there are certain outflowing water components coming into contact with the soil for a very short time. SiO_2, in particular, is a substance practically absent in rain water. The runoff of rain water on the surface of the ground will therefore lower the SiO_2 concentration in the surface water. This can also interpret the fall in pH and SiO_2 concentration measured at the spring outlets in the period from the winter to the spring thaw season.

20. There are other measurement data suggesting the same possibility. The flow of the groundwater in the lateral axis direction is marked by water accumulation in points B or C on the A - J section. As a result, the end points of streams B and C are located not in the center of the wetland but more to the west, with the pool having developed in this area. Water samples were taken at the pool location 0 situated at the end point of stream C on a total of four occasions from August through December 1991. The analysis results are presented in the bottom part of Table 2. It is evident that the annual mean chemical data for the pool are lowered, as compared with the corresponding values for the Oshimizu spring D0 on all counts, that is, pH, SiO_2, Na and K. In terms of NH_4-N, Mg and Ca, however, the former two chemicals have higher levels than the Oshimizu spring. As a result, it may be concluded that the seasonal changes in water chemistry shown in Fig. 5 are due to the runoff of groundwater very close to the ground surface and coming briefly into contact with the soil and plants.

21. In this manner, the Miyatoko wetland groundwater at a depth 50cm and below has a long residence time as evidenced by the water chemistry. Even when it flows out, it is still downstream of the wetland's outlets so that it will not significantly influence the surface water in the wetland zone. In contrast, however, the groundwater near the surface has been found to exhibit very clear-cut seasonal changes in pH, with the possibility that it may be very rapidly replaced through rainwater infiltration and outflow.

REFERENCES

1. HAMON W.R. Estimating potential evapotranspiration. ASCE HY3, paper 2817, 1961.
2. HIRATA T. and MURAOKA K. The relation between water migration and chemical processes in a forest ecosystem. IAHS Publ. No 215, 31-40, 1993.
3. HANYA T. et al. Chemical Environment of Earth. Kyouritsu Publ. 368p, 1978 (in Japanese).

The management of lagoons to conserve their natural heritage

Dr J. G. MUNFORD, Scottish Natural Heritage, and
Dr D. LAFFOLEY, English Nature

SYNOPSIS. The importance of lagoons as a habitat for wildlife has been recognised by the European Community through Directive 92/43/EEC. Lagoons are often seen as prime areas for land claim. This, coupled to the natural infilling of many lagoons, has led to an unacceptable reduction of this habitat in Great Britain. The lagoons of Italy have been largely safeguarded by the evolution of economic activity, which protects them from tourism based development evident along much of the remaining coastline. The mix of extensive lagoon based aquaculture, wild-fowling and the historical cultural significance of these wetlands offer lessons that are applicable to the management and safeguard of some British lagoons.

INTRODUCTION

1. Lagoons are bodies of saline water separated from the sea typically by low sand banks, or shingle, such as to restrict the ingress of sea water. The separation from the sea may be complete, as in the case of the Fleet separated from the sea by Chesil beach, or there may be restricted connections to the sea. They are primarily a feature of a submerging coast. The essence of a lagoon is the limited water exchange with the sea and the expanse of saline water which becomes increasingly land locked. Within this definition of a lagoon it is possible to include even features of emerging coastlines. The definition includes man made features such as the Broads which share many of the features of the large natural lagoons found in the Mediterranean. It also can apply to the variously named obs, houbs and vadills of the Scottish coast and which have their origin in the glacial activity of the last ice age.

2. These areas often have a cultural significance which far exceeds their geographic extent. Large lagoons are often highly productive biologically and not unnaturally have always been esteemed for fishing. Their geomorphology has lent itself to early attempts at aquaculture or fish trapping (ref.1). This paper examines the role of aquaculture in

ensuring the conservation of lagoons and how best this might be achieved.

REVIEW OF PUBLISHED KNOWLEDGE ON ENGLISH LAGOONS

3. Most of the knowledge of saline lagoons in England arises from studies conducted by the Nature Conservancy Council between 1984 and 1989. Little was known at that time about the extent or biology of saline lagoons or their conservation importance. During the studies 444 potential sites were visited of which 210 were surveyed, the remainder having been found not to contain brackish water or to have been infilled (ref. 2).

4. These studies showed that the main concentrations of saline lagoons occur around the south and east of England. Such a distribution might be expected due to the prevalence of low lying land bordering estuaries in these areas. Over 83% of lagoons in England can be found in close association with such sites (ref. 2). Smith and Laffoley (ref. 3) calculated that the total saline resource in England covers just 770 hectares (excluding the Fleet) and that the majority of saline lagoons are of less than 10 hectares area.

5. Barnes (ref. 4) and Sheader & Sheader (ref. 5) classified the lagoon resource as consisting either of natural lagoons or man made structures and evaluated their conservation status (Table 1). Both types of structure can support plant and animal species characteristic of saline lagoon conditions. Barnes identified that there are only 41 'natural' lagoons in England with a total area of 660 hectares.

6. Sheader & Sheader (ref. 5) divided the man made lagoons into four types: sea inlets, where seawater enters on each tide and no sill to retain water at low tide; isolated lagoons, where the lagoon is isolated from the sea by saltmarsh or seawalls and seawater enters by overtopping or ground water percolation; percolation pools in shingle banks; and silled or sluiced ponds where water is retained at all stages of the tide. Davidson *et al* (ref.2) concluded that of these types the most prevalent are silled ponds and that percolation pools and sea inlet sites are the rarest. Man made lagoons are also the most frequent type of lagoon around England, with some 75% of sites formed as a result of mans' influence. Indeed Barnes (ref. 6) concluded that on the north sea coast only three out of the 26 brackish lagoons he examined were natural in origin.

Table 1. Lagoons and lagoon like habitats recommended for conservation (after Smith & Laffoley 1992).

Lagoon name	County	Conservation status
Cresswell Ponds	Northumberland	SSSI
Easington Lagoons	N. Humberside	SSSI
Snettisham	Norfolk	None
Heacham Harbour	Norfolk	None
Broadwater	Norfolk	SSSI
Holkham Salts Hole	Norfolk	SSSI
Abraham's Bosom	Norfolk	SSSI
Blakeney Spit Pools	Norfolk	SSSI
The Denes	Suffolk	NNR
Benacre Broad	Suffolk	SSSI
Covehithe Broad	Suffolk	NNR
Shingle Street	Suffolk	SSSI
Aldeburgh P8	Suffolk	SSSI
Widewater	W. Sussex	SNCI
Rye Harbour	E. Sussex	SSSI
The Fleet	Dorset	SSSI
Swanpool	Cornwall	None
Lagoon like habitats		
Killingholme Pools	S. Humberside	None [1]
Humberston Fitties	Lincolnshire	SSSI
Cliffe Marshes	N. Kent	SSSI
Pagham lagoons	W. Sussex	SSSI
Birdham Pool	W. Sussex	None
Shut Lake	Hampshire	SSSI
Keyhaven-Lymington	Hampshire	SSSI
Gilkicker	Hampshire	SSSI
Little Anglesey	Hampshire	SSSI
Brading Marshes	IoW	SSSI [2]
Bembridge	IoW	SSSI [2]
Horsey Island	Devon	None

[1] Site owned by National Power
[2] Lagoons not mentioned in SSSI citation and therefore not afforded direct protection

ISOLATED SALINE WATERS SURVEY

7. The lagoons of Scotland have not received the same attention as those in England. In all, some 18 reports were produced by the now defunct Nature Conservancy Council on lagoons or their Scottish equivalents. This survey work is being continued by the Survey of Isolated Saline Waters in Scotland which will collect data on approximately 170 individual lagoons. To date these studies have confirmed the importance of lagoons to wildlife in Britain.

8. In particular they are extending our understanding of the importance of those enclosed bodies of saline waters such as those glacial features of Scotland, which have not previously been recognised as sharing the same features as lagoons of submerging coasts.

NATURAL HERITAGE INTEREST

9. Barnes (ref. 4) identified 38 specialist lagoonal species in Britain (Table 2). These are species defined as distinctly more characteristic of lagoon-like habitats than fresh water, estuarine, brackish water or the sea. Seven of these species are protected under Schedule 5, and one under Schedule 8, of the Wildlife and Countryside Act 1981. Of these, Ivell's sea anemone, *Edwardsia ivelli*, is only known from Widewater Lagoon. It has not been seen at this site since 1983. An extensive search in 1990 failed to find any individuals and it is now thought that this species is extinct. The Lagoon Sand Worm, *Armandia cirrhosa*, is also known from a single site, Eight Acre Pond near Lymington, and has not been seen since 1987. The only protected lagoon plant, the foxtail stonewort *Lamprothamnion papulosum* is also restricted to a sites around the Solent. Other species such as the lagoon shrimp *Gammarus insensibilis* have a more extensive distribution, inhabiting lagoons from the Solent as far around as the Humber on the North Sea coast.

10. Further information on lagoon species has been obtained by Bamber *et al* (ref. 7) who analysed habitat and animal and plant communities for 166 of the sites investigated in the NCC studies. Statistical analysis of data showed that six suites of species existed: freshwater species; stenohaline marine lagoon specialists; euryhaline lagoon specialists; estuarine species tolerant of lagoons; estuarine species incidental in lagoons; and under-recorded species. Bamber *et al* (ref. 7) showed that the best lagoons for specialist species were sites that were either bar built or sluiced lagoons with channel inlets and salinity close to 35‰. Exchange of water is also important and sites with high exchanges, of up to 40% volume, produce optimum conditions. In these situations the specialist lagoon species are able to out-compete other species to dominate the fauna and flora.

Table 2. Specialist British lagoonal plant and animal species (after Barnes 1988)

Plants	Mollusca (Molluscs)
Chara canescens	*Hydrobia ventrosa*
Chara baltica	*Hydrobia neglecta*
Chara connivens	*Onoba aculeus*
Lamprothamnion papulosum	*Littorina tenebrosa*
Tolypella n. nidifica	*Tenellia adspersa*
Ruppia maritima	*Cerastoderma glaucum*
Cnidaria (Anemones and hydroids)	Bryozoa (Bryozoans)
Gonothyraea loveni	*Conopeum seurati*
Edwardsia ivelli	*Victorella pavida*
Nematostella vectensis	
	Insecta (Insects)
	Sigara selecta
Polychaeta (Polychaete worms)	*Sigara stagnalis*
	Sigara coccina
Armandia cirrhosa	*Agabus conspersus*
Alkmaria romijni	*Berosus spinosus*
	Coelambus parallelogrammus
Crustacea (Crustaceans)	*Dytiscus circumflexus*
	Enochrus melanocephalus
Corophium insidiosum	*Enochrus bicolor*
Palaemonetes varians	*Enochrus haliphilus*
Sphaeroma hookeri	*Haliplus apicalis*
Idotea chelipes	*Ochthebius marinus*
Gammarus chevreuxi	*Paracymus aenus*

EC DIRECTIVE AND IMPLICATIONS

11. Barnes (ref. 4) and Sheader & Sheader (ref. 5) identified 29 sites which are the best examples of the particular types of lagoons in England. Almost half of these are protected as Sites of Special Scientific Interest under the 1981 Wildlife and Countryside Act. The conservation status of these sites will, however, receive a further level of protection due to the requirements of the EC Directive 92/43/EEC on the conservation of natural habitats and of wild fauna and flora.

12. This Directive, better known as the Habitats and Species Directive, will establish an international series of protected wildlife sites

in Europe by 2004. Under this Directive saline lagoons are identified as a priority habitat in recognition of the severe threats that they face. Only four other of the 22 listed coastal and halophytic habitats, have been given the same status. As a priority habitat they will be afforded a higher level of protection than other habitats. In particular it means that a site may only be affected by issues relating to human health or safety or other imperative reasons of overriding public interest. There is an agreed timetable for the Directive. Sites from Britain have to be proposed to the European Community for their consideration by June 1995.

13. In addition to the priority habitat status, saline lagoons may be particularly appropriate for consideration under the recovery aspect of this Directive. Lagoons by their nature are, in England, usually small decreet bodies of water and usually smaller than 10 hectares in area (Davidson et al 1991). A number have declined in quality over the past 50 years, such as Widewater lagoon in Sussex (Sheader - pers.comm) or Benacre Broad in Suffolk. It would, therefore, seem appropriate in certain circumstances for recovery programmes to be established on key sites. A first stage in this process will require the development of a management plan to scope the issues and identify options for management. For Widewater lagoon a management plan is already being developed (Patmore - pers.comm.) but options for reversing the decline in lagoon water quality and species diversity appear to be very expensive. Whitten (ref. 8) who examined the cost of recovering populations of lagoon species protected under the Wildlife and Countryside Act, estimates costs ranging from £24,750 up to £115,500. Given these high costs it is important to consider long term stability of sites before proceeding with any scheme. The costs are such that recovery work is too expensive for the statutory nature conservation bodies to fund as part of their regular duties and may therefore require supporting from the LIFE fund, developed to underpin this Directive. Such funds could also be used to support lagoon creation schemes as new lagoons should be relatively easy to establish in appropriate conditions.

RISKS AND THREATS

14. Long term stability of individual lagoons sites is another particular problem for the conservation of saline lagoons, whether through domestic legislation, such as SSSIs, or as part of the EC Habitats and Species Directive. By their very nature, saline lagoons form where the sea meets low lying land. In the north Atlantic coastal area, saline lagoons formed in macro-tidal conditions are fairly short lived features. In a natural system new lagoons will form as old lagoons are either lost to the sea by erosion, dry out, or turn into freshwater pools as saline influence is reduced.

MANAGEMENT SOLUTIONS

15. In England, natural lagoons are rare (ref. 4) and most form behind man made features such as costal and sea defence works. Those that form in natural situations are particularly susceptible to costal erosion and rising sea levels, combined in the south east of England, with sinking land mass resulting from the effects of last ice age. Often formation of new lagoons further inland is prevented by rising land levels or defences to protect improved or reclaimed agricultural land or population centres.

16. In situations where lagoons form behind man made features they may equally be under threat. In many places lagoons have been maintained due to the construction of the defence works, such as at Lymington in Hampshire. Sea defences, however, need repairing and replacing and in such situations improving the scale of the defences may obliterate the saline lagoons. At Lymington the lagoons were dominated by the highest known density of the protected lagoon anemone *Nematostella vectensis*. In this situation close cooperation between the National Rivers Authority, MAFF, English Nature and local interests resulted in a modified sea wall design which incorporated the lagoons (Tubbs pers comm).

RATE OF DISAPPEARANCE OF LAGOONS PARTICULARLY IN ENGLAND

17. The scale of loss of English saline lagoons can be calculated from data complied by Smith and Laffoley (ref. 3) who drew together records produced as a result of the NCC lagoon surveys in the 1980s. This work showed that between 1984 and 1989, 38 lagoons had been lost. The main cause was identified as infilling (31 sites) with the remaining sites having been lost to coastal erosion, sedimentation and drainage. This number of destroyed sites is worrying in scale and almost equates to the number of extant natural lagoons in England.

18. Further research in this area was commissioned by English Nature as part of the work to develop the concept of managed retreat. Managed retreat is a term used to describe situations where coastal defences, eroded by the sea, are not maintained. Instead, sea defences are constructed further inland and at a fraction of the cost. This also has the effect of creating new coastal habitats to stem the loss of such habitats to coastal erosion and sinking land and rising sea levels. Pye & French (ref. 9) looked at the loss of eight coastal habitats by examining information contained in databases, reports and published papers. By comparing the projected loss of habitats to their current extent they showed the highest level of loss is for saline lagoons where some 10% of the current resource may disappear by 2012 (Table 3).

19. Managed retreat may offer possibilities for the creation of new

saline lagoons, but opportunities for these coastal defence works are likely to be limited to situations where low grade agricultural land is available and high investment areas, such as towns and industrial complexes, are not affected.

20. The continuing pressures on saline lagoons from man-made or natural forces makes it particularly important to establish effective management on the existing key sites. To ensure the survival of specialist lagoon species which are restricted to only a limited number of sites it may be necessary to examine more radical options. These may include establishing 'insurance' populations is disused docks where they can be maintained in a man-made environment and reintroduction into 'natural' lagoons should the need arise.

Table 3. Approximate national loss of coastal habitats over the next 20 years resulting from combined natural and anthropogenic causes (Pye and French 1992).

Habitat	Area (ha)	% of existing resource
sand dunes	240	3%
saltmarsh	2750	8%
intertidal flats	10,000	4%
shingle formations	200	4%
saline lagoons	120	10%
soft cliff	10 km	4%
maritime cliff grassland	150	?
coastal heath	50	?

MANAGEMENT FOR CONSERVATION

21. Lagoons are subject to both natural and anthropogenic forces which ultimately limit their life, they are transient features in our landscape. Their conservation importance arising from the range of unique fauna and flora they sustain and the importance of wetland generally to migratory and mobile species including both fish and birds, is such that active management is required to protect this habitat. The processes which lead to the development of lagoons are actively being inhibited by man. It can be expected that the formation of new lagoons will not keep pace with the process of lagoon loss. Active management to protect existing lagoons is required to protect this important wildlife habitat. This may be achieved through a proscriptive approach if the site has been designated for its conservation importance. A more satisfactory approach is one of management whereby conservation importance is

protected, but which also allows for a degree of compatible use by man. It is possible already to see evidence of this form of mixed use. Recreational boating, angling and water sports can be compatible with conservation objectives if managed with conservation in mind. Aquaculture may also be compatible but problems can arise from the form of aquaculture practised. In general the more intensive the form of aquaculture the greater the risk of damage to the natural system.

22. The culture of fish ranges from extensive low intensity systems to modern intensive systems which concentrate production and make use of feed formulations and medicants to permit high concentrations of selected species to held in tanks or net cages with varying environmental control. The interaction between aquaculture and conservation has been discussed by Munford and Baxter (ref. 10) who conclude that there are ways in which can become compatible. Intrinsic to an understanding of how this might be achieved is the recognition that conservation is not necessarily a use of the environment requiring the identification and designation of specific sites but rather it can be seen as a sensible management consideration or objective for any form of environmental use. Only in exceptional cases will the proscriptive approach to conservation become necessary. In most cases the advocacy of good stewardship through sustainable use of resources and the precautionary principle should be adequate to safeguard or natural heritage (ref.11). There are few if any habitats which have not already been shaped to some extent by the man. In this past any safeguard of the natural resource has been co-incidental or resulted from man's inability to radically alter the natural system. We are now able to grossly affect our environment, but we also have a greater and increasing understanding of the underlying mechanisms which control natural systems which permit a more benign enlightened management.

THE USE OF LAGOONS FOR AQUACULTURE

23. The lagoons of the Adriatic coast of Italy have been used for aquaculture for many years (ref.12). It is likely that early attempts at fish farming were developed as long ago as during the Roman civilisation. There are now two different forms of aquaculture, one based on shellfish, largely the blue mussel Mytilus edulis, and the other more complex system of fish culture called valle culture in which various species of mullet are grown with gilt head bream, eel and sea bass (ref. 13). Shellfish culture relies on collection of wild seed and its subsequent cultivation on suspension systems to reduce losses to the mussel's principal predator, the common green crab *Carcinus maenus*. The fish cultivation system also relies heavily on fish fry caught in the wild although the modern evolution of this system has led to the hatchery rearing of the more expensive species. The original valle system is based on the observation that the fry of some marine species

have their early development within the relative shelter of saltmarshes and shallow lagoons. This reduces the predatory pressure to be expected in the open sea, whilst taking advantage of the high productivity of these areas. The temperature elevation evident during the spring, summer and autumn sustains a higher growth rate than would be possible at sea. In the natural system fish would return to sea for their first winter to escape the very low temperatures of these shallow areas, but they return again in spring to complete their development before breeding at sea. The system of aquaculture exploits these natural rhythms.

24. The valle consists of a series of basins enclosed by dykes. Usually there are four basins of varying construction. A small basin is used for the smallest fry. The water in this small basin becomes green through algal growth and is suitable for cultivating the small fish. Later in their development the fish are transferred through sluice gates to the first rearing pond which is usually of many hectares. At the end of the first autumn the complex design features of the valle first become evident. The whole complex of basins is surrounded by two canals, one of fresh water and one of sea water. The introduction of either fresh or salt water is used to call the fish. Thus, at the end of the first year, salt water is introduced to call the fish to the winter holding basin. The fish, which would naturally be returning to the open sea, follow the salinity gradient until they enter the over-wintering basin. This is the most carefully constructed feature of the whole system. It is basically of herring bone design. The sides of the enclosing dykes are as steep as possible and there are small trees and bushes along the upper margins. The herring bone is orientated to minimise wind waves on the surface of the side arms of the herring bone. The steep banks and planting also serve to reduce the effect of the wind. As the temperature drops, fresh water is flooded into the over-wintering system through stand pipes. The fresh water pools on the cold dense saline water and subsequently freezes. This frozen surface water acts in much the same way as a glass house and can maintain the underlying saline water at temperatures over 5°C. This is important as the species in cultivation die below this temperature. Modern valle culture also uses large boiler houses to heat the underlying saline water to reduce mortalities to a minimum. Following the first winter the fish are called using fresh water to the second large basin to complete their development to market size.

25. Harvesting also relies on the natural migratory behaviour of the fish. In this case introduction of saline water calls the fish to a system of grids which both harvest and sort the fish by size. Modern systems are based on refinements both in design and materials on the early use of rush and brushwood barriers designed to separate the largest species, such as mullet, from the smaller gilt head bream and eels.

MANAGEMENT SOLUTIONS

26. The importance of this system to this part of Italy is that it has given a value to these lagoon areas, which has protected them from the intensive tourist orientated development evident along so much of the remaining coastline. The value of the valle system of cultivation is largely cultural. The dependence on this source of fish protein which must have been of vital importance to the local population in the past has disappeared with modern agricultural and fisheries development, but the cultural significance is still very evident. The principal harvesting takes place before Christmas. The various species of mullet form as much as part of the tradition fare at this time of year as turkey does in Britain. There is also a economic value to the valle. In strict economic terms it is unlikely that the return per hectare from fish cultivation is viable. Attempts have been made to increase the return, largely through cultivation of the highly priced sea bass, but this has led in some cases to the use of artificial feeds with disastrous consequences on the health of the lagoons. The whole system of valle culture is intimately linked with the natural productivity of the lagoons coupled with minimising exchange to increase the rate of temperature increase. It is not possible to reconcile this with increased productivity through the addition of fish feed. The extra biological load arising from waste food and fish excretion can unbalance the system, although increased aeration can help.

27. The problem of pollution is very evident in the management of the valle. In some cases owners prohibit even the use of outboard motors to reduce the risk of pollution.

28. The main profit arising from this system of fish culture is derived from a fortuitous natural bonus. The protection of what are in essence wet lands has led to this habitat being of vital importance to wildfowl. The valle owners derive their principal profit (and in some cases personal pleasure) from the wild-fowling which takes place on the two main cultivation basins. These shallow, but extensive, expanses of water have old wine butts at strategic locations to provide shooting hides. Suitably luxurious lodges are also provided to encourage use of the facilities by shooting parties over a number of days.

29. It is evident that this system of wetland management has afforded some protection to this important habitat in an area where the economic pressures are such that land claim and development might have been expected. An important aspect of this system is that the lagoons are privately owned and that the owners perpetuate the system of management not only for the historic reasons of fish cultivation, but also for pleasure and recreation. Shellfish culture (ref. 14), which is also carried out in lagoonal areas, shows an alterative development that also gives the lagoons an economic and cultural value However, it may

suffer because of the nature of public ownership, which is not always amenable to strict control. Public ownership has resulted in planning decisions being taken that, whilst of benefit to the overall social good, have resulted in damage and degrading of the lagoonal habitat. An example can be seen at the southern margin of the Venice lagoon where improvements to road alinement has resulted in the construction of a supporting causeway without adequate provision of culverts. The enclosed area of the lagoon has silted up increasing markedly the natural reversion of such areas to low grade land. The pressure to develop this land in some way will inevitably result in a loss to the wetland habitat around the lagoon margin. Public control of the Venice lagoon has also been unable to minimise the impact of industrial development on the lagoon margin at Mestre.

30. Of more immediate relevance to this discussion is the control and environmental impact of the shellfish farming evident in this lagoon and others in public control, notably the lagoons at Comacchio and Grado. Of all systems of aquaculture, shellfish farming appears to be least damaging to the environment (ref. 10). The various system ultimately rely on the natural productivity of sea water. All the species of bivalve molluscs (clams, mussels and oysters) are filter feeders relying for sustenance on the naturally occurring microscopic phytoplankton. Shallow sea areas are naturally productive, with the lagoons also being subjected to agricultural run off and thus increased fertilisation. Shellfish culture can in fact reduce the worst effects of the enhanced plant production arising from agricultural fertilisers and urban sewage but the Venice lagoon in particular is severely affected by increased plant production and subsequent anoxia during breakdown of this material resulting in extensive fish deaths. The same effect is also evident in other lagoons.

31. There is a negative aspect to shellfish culture in lagoons with restricted water flow. All shellfish deposit silt. This arises from the filtering activity of the shellfish and through the deposition of faecal material. In the Venice lagoon this may result in up to one metre of silt being deposited annually beneath shellfish cultivation areas. Ultimately the sites become untenable.

CONCLUSIONS

32. It is evident that aquaculture can be compatible in certain circumstances with the requirements of conservation indeed the successful prosecution of aquaculture made only be possible if environmental safeguards are built into its overall management structure. Economic value and cultural value can become driving forces for habitat conservation. At least in the case of the former, it is possible to make comparisons with proposed development schemes, but there are very

real dangers in trying to judge an area on economic grounds alone. Conservation should not necessarily be seen as proscriptive. Whenever possible it is better for conservation to be seen as a management objective and to permit the public at large to benefit from the protected environment. The value of Britain's wildlife is becoming increasingly appreciated it is to be hoped that Britain's lagoons will also be recognised as a valuable resource and be afforded the degree of protection they require either as completely protected sites or through enlightened management of their economic potential.

REFERENCES
1. d'Ancona U. Fishing and fish culture in brackish-water lagoons. Fisheries Bulletin FAO,1954, vol.7, 147-172.
2. Davidson, N.C. et al. Nature Conservation and estuaries in Great Britain. Nature Conservancy Council, Peterborough, 1991.
3. Smith, B.P. & Laffoley, D. d'A. A directory of saline lagoons and lagoon like habitats in England. English Nature, Peterborough, 1992.
4. Barnes, R.S.K. The coastal lagoons of Britain: an overview. Nature Conservancy Council, 1988, CSD Report, No.933.
5. Sheader, M & Sheader, A. The coastal saline ponds of England and Wales: an overview. Nature Conservancy Council, 1989, CSD Report, No.1009.
6. Barnes, R.S.K. The coastal lagoons of East Anglia, U.K. Journal of Coastal Research, 1987, vol. 3, 417-427.
7. Bamber, R.N. et al. On the ecology of brackish water lagoons in Great Britain. Aquatic Conservation, 1992, vol.2 , 65-94.
8. Whitten, A.J. Recovery: A proposed programme for Britain's protected species. Nature Conservancy Council, 1990, CSD Report, No.1089.
9. Pye, K. & French, P.W. Targets for coastal habitat recreation. Peterborough, English Nature, 1992.
10. Munford J.G. and Baxter J.M. Conservation and Aquaculture. Aquaculture and the Environment,. European Aquaculture Society Special Publication ,1991, No.16, 279-298.
11. Munford J.G. and Donnan D. W. The sustainable use of the renewable marine resources of the Scottish islands. The Islands of Scotland: A living marine heritage. HMSO, Edinburgh. 1994.
12. de Angelis R. Brackish-water lagoons and their exploitation. General Fisheries Council for the Mediterranean. 1960, Studies and Reviews, No. 12.
13. Zerbinato M. L'acquicoltura in Italia: indagine conoscitiva. Associazione Nazionale Giovani Agricoltori, Rome. 1981, 126p..
14. Munford J.G. et al. A study on the mass mortality of mussels in the Laguna Veneta. Journal of the World Mariculture Society. 1981, vol.12, 186-199.

An integrated nature conservation and development strategy for the Deep Bay wetlands, Yuen Long, Hong Kong

G. GRANT, BSc(Hons), Senior Ecologist, London Conservation Services

SYNOPSIS. EcoSchemes (Asia) Ltd has been instructed by leading Hong Kong developers to identify key nature conservation objectives for the fishponds of the Deep Bay area of Hong Kong and to formulate practical plans to integrate these objectives into a property development strategy which can lead to establishment of secure and sympathetic land use for the long term.

THE STUDY AREA
1. Our work concentrated on land use and nature conservation issues in the low lying area of approximately 1000 hectares which lies to the south of Deep Bay in the North West New Territories of Hong Kong. The border between Hong Kong and China bisects Deep Bay. The study area was to the immediate north of Yuen Long, an area bounded in the west by the Lau Fau Shan (Tsim Bei Tsui) peninsula and to the east by the Castle Peak Road and Lok Ma Chau border crossing.

A BRIEF DESCRIPTION AND HISTORY OF THE AREA
2. A hundred years ago, the low lying areas of Deep Bay were almost entirely covered by intertidal mud and mangroves. Over the centuries some small scale reclamation was probably undertaken, but it was not until this century that large scale reclamation took place, with the establishment of brackish water rice paddies and shrimp ponds (gei wai). By 1945 about half of the former area of mangrove had been converted to gei wai, but the intertidal nature of the ecosystem was still largely intact, since gei wai retain many of the ecological characteristics of the mangroves they are formed from.
3. In the post war period both brackish water rice paddies and gei wai were replaced by freshwater fish ponds. In each case, the change from gei wai to freshwater fishpond is fundamental and dramatic, for the tide is completely excluded and mangroves are removed during construction. The destruction of wildlife habitat caused in this way was recognised by the Hong Kong Government, which declared, in 1976, the largest remaining area of mangrove and gei wai at Tai Long Kei and Lut Chau

as the Mai Po Marshes Site of Special Scientific Interest (SSSI). However the loss of gei wai and mangrove to fish ponds has continued.

4. The intertidal mud of Deep Bay remains, and is of international importance as a feeding and roosting area for wintering and passage birds. In Deep Bay, during the spring migration, typically 10,000 birds may be present at a time (the highest winter count was over 51,000 birds in mid-January 1992) and there has been a noticeable increase in numbers during the last decade - almost certainly as a result of habitat loss elsewhere in the Pearl River Estuary. The nature reserve at Mai Po complements Deep Bay by providing a secure high tide roost for many of these migrating birds, particularly waders.

5. Although they meet the wide ranging definition of wetlands provided by the Ramsar Convention (the international convention for the conservation of wetlands), the remaining freshwater fishponds are of little value for the vast majority of these migrating birds - they are too disturbed and the water is usually too deep. However they do provide part of the diet of resident fish eating species of birds such as egrets, herons and kingfishers.

6. A more recent change to the landscape (since the 1970s) has been the infilling of fishponds for industrial use (e.g. the Yuen Long Industrial Estate) or for residential development (e.g. Fairview Park, a private development and Tin Shui Wai, a government planned new town). As well as such large scale authorised infilling there has been localised illegal infilling of ponds in order to create open storage facilities and vehicle repair yards.

7. Another major consideration must be the continuing reduction in water quality of the rivers which issue into the bay and the water quality of Deep Bay itself. Shortly before it enters Deep Bay, the Kam Tin River, for example, has virtually no dissolved oxygen, very high levels of suspended solids and a water quality index of 15 (Very Bad). Most of this pollution emanates from livestock farms upstream. Although Deep Bay is naturally a high nutrient ecosystem (with large quantities of organic material suspended in the tidal waters), it is not known at what point such an increase in pollution might have a damaging effect on the range of species and total numbers of individual invertebrates which occur in the mudflats. It must be emphasised that it is these mud-dwelling invertebrates which support the vast flocks of birds and although some invertebrates are especially tolerant of pollution they would not support many of the birds with highly specialised feeding habits that currently occur.

8. Also of concern is the pollution from industrial sources, for example from existing and planned industrial estates in the Special Economic Zone of Shenzhen (just across the border in Guangdong Province, China), which may accumulate in the mud, reaching critical levels and causing the sudden collapse of populations of susceptible groups and

species of invertebrates and those birds which rely on them. The possibility of such a catastrophic event should give reason for caution in increasing the pollution load on Deep Bay, particularly industrial pollutants.

CURRENT TRENDS

9. Agriculture and fishing continue to decline as people find better paid work in nearby factories and offices. Fish farming and agriculture is often a part time occupation, increasingly characterised by the use of modern techniques (e.g. tracked excavators are used to rebuild ponds each year) and supplemented by light industrial activity (e.g. vehicle repair). The semi-official nature of much of this activity and a lack of overall management by the government means that domestic refuse is not collected and scrap equipment and vehicles are simply abandoned. The overall effect of this is that the environment of actively worked fishponds is deteriorating.

10. The increased susceptibility of the area to flooding (partly caused by the original reclamation to create fishponds, but also by urban development) has caused both the Hong Kong and Shenzhen Governments to propose or consider large scale modification of the river systems which feed Deep Bay.

11. River training schemes for the Kam Tin River and Yuen Long Nullah (two of the main rivers which enter the Bay from Hong Kong), as originally proposed, would have resulted in the loss of large areas of mangrove and would also have necessitated the dredging of inter tidal mud. (Large scale dredging in inner Deep Bay would not only result in the direct loss of feeding habitat for birds, but could also cause the release of pollutants from the deeper mud. These pollutants could then affect the invertebrates of an even larger area of mud flats than those directly affected by the dredging. Further maintenance dredging could prolong these effects indefinitely.) Following discussions with World Wide Fund for Nature Hong Kong, consultants acting for developers were able to propose an alternative scheme which avoided large scale clearance of mangrove and the dredging of intertidal mud. In addition, in the alternative scheme, provision was made for the planting of mangroves alongside newly created channels. This scheme has now been accepted by Government in a slightly modified form and work is now under way. It will dramatically change the landscape, with the main channel of the Kam Tin River Floodway being over 200m wide.

POLICY BACKGROUND

12. **Ramsar Agreement** (Convention on Wetlands of International Importance especially as Waterfowl Habitat). Hong Kong is party to this international convention through the United Kingdom, which is a signatory (The People's Republic of China is also a signatory). The

Hong Kong government has indicated that in principle Deep Bay can be considered suitable for designation and has set up a working group to consider the full implications of designation.

13. A recent meeting of the contracting parties of the Ramsar Agreement (Japan 1993) developed guidelines on the "wise use" of wetlands. The meeting noted that the case studies scrutinised by its wise use working group during the previous 3 years had led to the conclusion that "Social and economic factors are the main reasons for wetland loss and therefore need to be of central concern in wise use programmes." In introducing "Additional Guidance for the implementation of the wise use concept" it is stated that "In the early years of the Convention, the wise use provision proved to be difficult to apply" and goes on "Over time, as the essential need to integrate conservation and development has become recognised throughout the world, the contracting parties to the Ramsar Convention have made wise use a central theme for the functioning of the Convention."

14. Under "Action at particular wetland sites" the guidelines state that "In order to achieve wise use of wetlands, it is necessary to attain a balance that ensures the maintenance of all wetland types through activities that can range from strict protection all the way to active intervention, including restoration." This is particularly relevant in Deep Bay, where freshwater fishponds established on former mangrove will need modification if they are to fully meet the habitat requirements of migrating birds.

15. **Bonn Convention** (Convention on the Conservation of Migratory Species of Wild Animals). This international agreement is of particular relevance to Deep Bay given its importance as a stop off point for migrating shorebirds using the Siberian - Australasian flyway.

16. **Threatened species.** Deep Bay is visited by a number of birds listed as threatened or near threatened by the BirdLife International/IUCN Red Data Books. Deep Bay is of world importance (i.e. regularly supports more than 1% of the world population) for the Black-faced Spoonbill, Asiatic Dowitcher, Spotted Greenshank and Saunder's Gull.

17. **Buffer Zones** have been established by government as a means of giving added protection to Deep Bay through control of incompatible land use. Nature conservation is the primary planning intention for the inner Buffer Zone which adjoined Deep Bay. Development in the outer Buffer Zone should have "an insignificant impact on the environment, ecology, drainage, sewerage and traffic in the Inner Deep Bay wetland ecosystem". The Buffer Zone system does appear to assume that all change is likely to be negative and has not, in our view, taken proper account of the future need to rehabilitate the area, particularly areas covered by freshwater fishponds.

THE CONCEPT OF AN INTEGRATED NATURE CONSERVATION AND DEVELOPMENT STRATEGY

18. The integration of nature conservation with economic objectives is now widely promoted. Few societies are willing to set aside and protect from economic exploitation all areas of nature conservation interest and to provide the necessary revenue for long term management. The Hong Kong government, like most other governments, spends relatively small sums on nature conservation. (In 1993 the government agreed to spend HK$16m (£ 1.4m) on the resumption of Gei Wais from fish farmers at the Mai Po Marshes, an area to be included within the nature reserve managed by the World Wide Fund for Nature Hong Kong.)

19. In the Deep Bay area, government has tolerated land uses (namely freshwater fish farming and duck rearing) which are largely incompatible with key nature conservation objectives and which provide no income for those areas managed specifically for nature conservation (i.e. the Mai Po Nature Reserve). Furthermore, the recent decline in fish farming has led to even more damaging uses, which include infilling of ponds for vehicle and equipment storage.

20. The government's policy of development control has not been effective. Confrontations between government and landowners over the future of the area waste time and money and are unlikely to bring real benefits for nature conservation since development control is not enough in itself to prevent continuing decline. In contract, our approach has been to promote a partnership between government, business and the voluntary sector - an approach which involves some compromise on all sides, but which promises much. The objective is to replace existing incompatible and declining land uses with housing and recreational developments which will in turn provide the necessary funds for the rehabilitation and best possible management for wetland in the whole of the inner Buffer Zone and parts of the outer Buffer Zone and beyond. It is our belief that private property development (an exceptionally strong industry in Hong Kong) is the only realistic source of funding for the large scale, multi-billion dollar rehabilitation and management programme we envisage.

MAJOR OBJECTIVES OF OUR STRATEGY

21. Our major objectives were

(a) to provide suitable habitat for migratory birds
(b) to increase the overall area of mangrove - a habitat which once dominated the whole wetland
(c) to create new wetlands (including freshwater marsh) which will be managed to allow them to develop their maximum potential biological diversity

MANAGEMENT SOLUTIONS

(d) to provide a secure, properly funded form of management for these areas
(e) to improve water quality in Deep Bay and its catchment
(f) to continue to educate people, children and adults, both formally and informally on the importance of Deep Bay and of the enjoyment to be had through the study and observation of its wildlife
(g) to promote and encourage further cross border co-operation on all of the above objectives

HOW THE STRATEGY WILL WORK IN PRACTICE

22. **Fung Lok Wai.** EcoSchemes (Asia) Ltd has prepared a detailed plan for the creation of a nature reserve on 40 hectares of this area of fishponds to the west of the mouth of the Shan Pui River. The plan includes the creation of scrapes (high tide feeding and roosting areas) for shorebirds and waterfowl, freshwater marsh, reed beds and ponds. Much of the site will be accessible or viewable for educational purposes and there will be a visitor centre. The developer of the adjacent low rise residential development will provide the necessary funding for construction and will provide an endowment to meet the estimated HK$3m annual running costs for the nature reserve.

23. **Nam Sang Wai and Lut Chau.** EcoSchemes (Asia) Ltd has prepared a similar nature reserve scheme for the 50 hectare island of Lut Chau, on the eastern side of the mouth of the Shan Pui River. The developer seeking permission to build homes and a golf course on nearby ponds at Nam Sang Wai has offered to pay for the construction and long term management of a nature reserve at Lut Chau. Funds for managing the nature reserve will be provided, in this case, through an inflation-proof levy paid by future residents. Other off-site mitigation includes two river cleaning and screening plants designed to bring some rapid improvement to the grossly polluted rivers which flow by these sites. Again, as with Fung Lok Wai, this development complies with our recommended pattern for the integration of nature conservation and development.

WATER QUALITY

24. Declining water quality in Deep Bay and the associated potential for a catastrophic collapse in invertebrates and, in turn, bird populations means that action is urgently required to combat this problem. Already the managers of the Mai Po Nature Reserve have begun to isolate some of the tidal ponds in order to prevent them from becoming polluted. This has led to a reduction in salinity with a resulting decline in mangrove and increase in aggressive freshwater reeds, which are of less nature conservation value. The government does have a livestock waste control law, but enforcement has yet to take effect.

CONCLUSION

25. Urban development has been considered, by some people, as the major problem for the Deep Bay wetlands, but other factors, like pollution pose more serious threats. Much of the damage to this ecosystem was done when mangroves were turned into fish ponds earlier this century and there is an opportunity to restore and manage these degraded wetlands using money generated through property development. The problem is about to become part of the solution.

REFERENCE

1. IRVING R. and MORTON B. A Geography of the Mai Po Marshes. Hong Kong University Press/World Wide Fund for Nature, Hong Kong, 1988.